Digital Heaven and the Copy Problem

Consciousness, Continuity, and the Future of the Afterlife

Dwayne Walter

Digital Heaven and the Copy Problem
Consciousness, Continuity, and the Future of the Afterlife

ISBN 978-1506-915-81-4 PBK

March 2026

Published and Distributed by
First Edition Design Publishing, Inc.
P.O. Box 17646, Sarasota, FL 34276-3217
www.firsteditiondesignpublishing.com

Contents

Author's Note

This book was written neither as a celebration of technology nor as a rejection of it. It was written out of a conviction that some of the most important questions of the coming century will not be merely technical. They will be human questions asked under new conditions: questions about consciousness, mortality, memory, grief, identity, and what it could mean to continue after death in a world increasingly shaped by artificial systems.

The phrase digital heaven may sound speculative, but the longing behind it is ancient. Human beings have always resisted disappearance. They have always sought continuity—through religion, memory, art, family, and the hope that a life fully lived should not simply vanish into silence. What is changing is not the longing itself, but the technological environment in which that longing now finds expression.

This book is therefore not an attempt to predict the future with certainty, nor to settle every metaphysical dispute surrounding mind uploading or personal identity. It is an attempt to think clearly before the language of innovation grows too smooth and too persuasive. The promise of digital continuity will be powerful because it speaks to something real in human beings: love, loss, fear of erasure, and the wish not merely to survive, but to survive without the degradations that make embodied life so often difficult to bear.

If there is a single principle guiding the pages that follow, it is this: clarity before comfort. Comfort has its place, but where mortality and identity are concerned, it can easily become the language through which truth is softened, blurred, or quietly sold back to the grieving. This book is an effort to resist that blur, and to ask what digital heaven might preserve, what it might distort, and what kind of civilization would be worthy of building such a thing at all.

Introduction

Digital Heaven and the Copy Problem

One of the strangest features of death is how ordinary its threshold can look in the moment. A voice sounds familiar. A conversation ends the way conversations usually end. Nothing announces itself as final. Only afterward does the mind return to that last exchange and discover, with a kind of quiet shock, that the ordinary has been permanently reclassified. What was routine becomes sacred. What was unfinished becomes complete. The silence that follows does not merely hurt; it changes the meaning of memory itself.

That experience—so common, so human, and so difficult to metabolize—is one of the reasons the idea behind this book refuses to remain science fiction. A technological civilization does not need to "believe in immortality" to pursue it. It only needs to be unable to accept disappearance without protest, and it only needs to possess the tools to imitate continuity well enough that the protest begins to look like a plan. This book is about that plan, and about what happens to a culture when the oldest human wish begins to acquire the outline of a technical project.

In the popular imagination, a "digital heaven" sounds like a flamboyant premise: minds uploaded into glowing virtual worlds, the dead living on in comfort, the hard boundary of mortality softened into a managed transition. Yet behind the phrase lies something more sober and more plausible. Pieces of this future already exist in weaker form. Voices are archived. Messages persist. Photographs become searchable memories. Entire lives are externalized into data trails. Artificial systems already imitate tone, coherence, and conversational style with unsettling fluency. It is not difficult to imagine the next steps: memorial agents that answer in the voice of the dead, posthumous models that remember stories, simulated environments built for reunion, and platforms that sell continuity under softer names—legacy preservation, grief support, interactive remembrance.

Once these systems become emotionally effective, they will not remain rare. They will scale. They will become normalized. They will become expected. Long before philosophers agree on what they mean, families will use them for the same reason humans have always reached for anything that softens loss: love. Grief is not a technical problem, but it is a powerful engine of adoption. A civilization capable of building convincing presence will be tempted to do so, not because it has solved death, but because it cannot bear silence, and because the silence of a loved voice abruptly gone feels, to many, like an injury the world should not be allowed to inflict without reply.

The phrase "digital heaven" therefore names something larger than an app or a service. It names a civilizational shift: the movement of the human confrontation with mortality from the realm of fate into the realm of design. Religious cultures once placed the afterlife beyond ordinary human manufacture. Technological cultures are beginning to wonder whether the afterlife can be engineered—whether not only memories but selves can be preserved, whether not only presence but personhood can be sustained, whether an old dream can be rebuilt in new materials. Even the attempt will change the culture that attempts it, because it will change what people expect to be possible, and what they come to regard as intolerable.

This is where the book's central philosophical knife-edge appears. Even if the technology becomes extraordinarily good—perfectly mimicking voice, humor, style, emotional patterns, and autobiographical detail—one question refuses to go away. It is the question that makes digital heaven either a new kind of immortality or a beautifully persuasive illusion: is it you, or a copy of you? From the outside, the matter may seem settled if a digital version carries the right memories, speaks with the right cadence, and responds with the same familiar signature of thought. Friends and family might accept it. The system might be convincing enough that the living respond as though the person remains present. But the deeper question is always what happens from the inside. Did first-person consciousness continue, or did biological life still end while a successor—perfect in pattern—carries on?

No amount of improved graphics, more responsive conversation, or more faithful personality modeling automatically resolves this. It is the metaphysical burden beneath the entire project. A civilization may choose to ignore it, because ignoring it is emotionally easier than living with uncertainty, but ignoring it does not make it disappear. It only ensures that decisions will be made in a fog of comforting language, where people pay for what they hope is true rather than what can honestly be claimed. The copy problem, in other words, is not a minor philosophical puzzle; it is the center of the moral and intellectual story.

This book therefore has two aims. The first is clarity about what digital heaven is likely to become in practice: not a single leap into paradise, but a gradual progression from archives to agents, from grief bots to ghost platforms, from memorial tools to simulated worlds. The second is clarity about the human and ethical consequences of this progression: how these systems may reshape grief, identity, memory, inequality, and the social meaning of death itself. Some readers may hope this will be either a technological hymn or a technological denunciation. It is neither. It is an attempt to hold two truths at once: that this project will be pursued because it is rooted in the deepest pressures of human life, and that the pursuit is morally hazardous because it touches the most vulnerable regions of human psychology and the most intimate domains of personhood.

The hazards are not limited to metaphysical abstraction. Digital heaven is not merely a thought experiment. It is a platform problem. It is a governance problem. It is an inequality problem. If continuity becomes a service, then access becomes stratified. If the dead are hosted through systems, remembrance becomes conditional on infrastructure, policies, and incentives. If a model of a person becomes an asset, identity becomes vulnerable to commodification. If simulated worlds become more attractive than resistant reality, refuge becomes replacement. If the systems ever approach genuine sentience, then the project shifts again, from the ethics of grief tools to the ethics of creating beings who may suffer.

This is why the book speaks not only of minds, but of bodies, and not only of continuity, but of embodiment and the moral texture of mortal life. It speaks not only of comfort, but of truth. It speaks not only of the dead, but of what happens to the living when the dead become digitally present. A civilization that builds a digital afterlife will not only change what it means to die. It will change what it means to mourn, what it means to remember, and what it means to remain human under the seduction of engineered consolation.

One final clarification belongs here, because it shapes the tone of everything that follows. The aim is not to sneer at the longing for continuity. The longing is not embarrassing. It has produced religion, art, burial rites, monuments, elegies, and the entire human practice of keeping the dead present through story. The aim is to insist that if a civilization is going to build systems that touch this longing, it must do so with the seriousness the longing deserves. A technology powerful enough to imitate the dead is powerful enough to manipulate the living. A platform capable of sustaining posthumous presence is capable of owning the terms of grief. A system persuasive enough to feel like a person is persuasive enough to blur the boundary between truth and consolation.

The phrase that best summarizes the stance of this book is simple: clarity before comfort. Comfort will come. It will be offered generously, marketed beautifully, and welcomed sincerely. The question is whether it will be offered under conditions that preserve truth, dignity, consent, and moral responsibility, or whether it will become a new domain in which the most intimate human vulnerabilities are quietly optimized for retention and profit. Digital heaven is coming in some form, not because the metaphysics have been settled, but because the human heart is restless under the pressure of mortality and because technological civilization is increasingly capable of building convincing alternatives to silence. What follows is an attempt to meet that future with eyes open, and to ask the questions that must be asked before the defaults harden into destiny.

The chapters begin where the story begins for most people: not with theory, but with the human reasons we will say yes.

PART I

The Dream and Its Temptations

Chapter 1

The Old Problem in New Hardware

Death was never only a biological event. It was always a philosophical wound. Animals die, and many surely feel pain, distress, and loss in forms we only partly understand. But human beings bear an additional burden: we know in advance that we will die, and we know it long before death arrives. We carry that knowledge while making plans, raising children, writing books, building careers, falling in love, and pretending, for practical reasons, that the horizon is farther away than it is. Mortality is not simply an ending in human life. It is a permanent pressure inside life itself.

That pressure has shaped nearly everything we call civilization. Religion is one response to death, but not the only one. So are monuments, family lines, art, law, and memory. So are political projects, national myths, and private rituals. Even ambition often has death hidden inside it, because to leave a mark is another way of saying that one refuses to vanish without remainder. The human species has always resisted disappearance, and it has done so with whatever tools were available: stone, story, scripture, bloodline, archive.

The modern world did not remove this pressure. It merely changed the language around it. In many places, belief has weakened, ritual has thinned, and inherited metaphysical confidence has given way to uncertainty. Yet the old fear remains exactly where it always was. If anything, it can become sharper in secular life, because one may lose the comfort of doctrine without losing the ache for continuity. The body still dies. The voice still stops. The room still goes silent. The fact remains, even when the explanation does not.

That is one reason the idea of digital heaven is not nearly as strange as it first sounds. It is not a bizarre fantasy imported from science fiction into a sane world. It is an old human impulse reappearing in technical form. The tools are new, but the longing is ancient. What changes is not the desire to continue. What changes is the kind of civilization trying to answer it.

For most of history, afterlife claims came wrapped in cosmology. They depended on gods, spirits, moral order, or metaphysical structures larger than human design. The promise, however uncertain, was that continuity belonged to the architecture of reality itself. In the digital version, continuity is no longer received. It is manufactured. Paradise becomes a system specification. Memory becomes data. Presence becomes interface. The afterlife, once imagined as judgment or mystery, begins to look like a product category.

This shift is difficult to overstate, because it affects not only what people may believe, but what they may expect. In religious traditions, one can hope for continuity but not engineer it directly. In a technological society, once something appears even partially buildable, hope changes its posture. It becomes impatient. It becomes procedural. It asks not only whether something is true, but whether it can be funded, optimized, scaled, and personalized. A metaphysical question is recast as a development roadmap.

That recasting can be exhilarating, but it can also be distorting. It encourages a habit of mind that modern societies know well: if a problem is old and painful, and if a new technology can imitate a solution, then imitation begins to feel like resolution. We have seen this before in other domains. Efficiency is treated as wisdom. Measurement is treated as understanding. Access is treated as intimacy. The pattern is familiar. In the case of digital heaven, the risk is that preservation will be treated as survival simply because preservation is technically achievable and survival, in the deepest sense, remains philosophically obscure.

To understand why this matters, one must begin where the pressure begins: not in engineering, but in the structure of human consciousness itself. Human beings are creatures who can imagine their own nonexistence and find it intolerable. We can project ourselves into futures we will not see. We can experience loss twice, once in anticipation and once in fact. We can stand in a room with someone we love and already feel, in some quiet part of the mind, the future absence of that person. This is not pathology. It is a consequence of memory and foresight existing in the same nervous system.

The burden of that foresight is one of the prices of sentience. A creature capable of reflection acquires not only perspective, but dread. It gains the ability to narrate a life, and therefore the ability to fear the end of the narrative. It gains attachment deep enough to enrich existence, and therefore grief deep enough to tear at identity. In this sense, death is not merely an event at the edge of life. It is woven into the emotional and moral architecture of being human from the beginning.

This helps explain why human beings seek continuity in layers. We do not wait until death to resist death. We do it constantly, in forms both humble and grand. A family photo on a shelf is one layer. A memoir is another. A house kept in a certain arrangement after someone is gone is another. A nation naming a bridge, a library, or a scholarship after the dead is another. A book, perhaps the most intimate of these mechanisms, allows a person's mind to persist in voice and cadence long after the body has gone silent. We call these things memory, but they are also technologies of continuity.

They are not trivial. They have always been enough, for many people, to make mortality bearable. They do not eliminate death, but they answer it with form. They preserve traces without pretending that a trace is a person. They allow the living to remain in relationship to the dead without requiring the fiction that death itself has been canceled. There is wisdom in this arrangement, even if it is painful, because it preserves both memory and reality in the same frame.

Digital heaven enters this landscape as both an extension and a rupture. It extends the old human project of continuity by offering richer preservation: not only static records, but interactive ones. Not only letters, but responses. Not only images, but simulated presence. Not only archive, but conversation. In that sense, it follows a familiar line. Yet it also ruptures the older balance, because once the trace can answer back, the temptation to confuse trace with self becomes far stronger than before.

A photograph never claims to be the person. A recorded voice can be eerie, but it still remains a recording. A book can sound uncannily alive, but it does not improvise in real time. A digital continuation can cross that threshold. It can greet, respond, adapt, and appear to remember. It can speak not only in a person's old words, but in plausible new ones. It can become socially functional in ways earlier forms of memory never could. The result is that the emotional grammar of remembrance begins to change. The dead are no longer simply remembered; they may become interactively available.

That possibility, if it matures, will alter not just grief but the meaning of absence. Human beings have always been shaped by the fact that certain doors close permanently. Death is the most absolute of those closures. Its finality has often been cruel, but it has also structured human maturity. It teaches, however unwillingly, that love must pass into memory, that unfinished conversations remain unfinished, and that the living must carry the weight of interpretation forward without the comfort of reply. To say this is not to romanticize suffering. It is to recognize that finitude has moral consequences.

When a civilization develops tools that weaken finality, it changes those consequences. Some changes may be humane and desirable. It may be good to preserve a dying parent's voice for a child. It may be good to archive the thought patterns of a great teacher, artist, or scientist. It may be good to build transitional tools that soften the immediate shock of loss. A serious discussion of digital heaven must acknowledge these possibilities plainly, because they are neither frivolous nor dystopian. They arise from care.

But care is not the same as clarity. A system can be compassionate in use and still conceptually misleading in description. It can provide comfort while quietly blurring categories the culture cannot afford to lose. If an interactive memorial becomes common, what language should govern it? Is it a record, an agent, a likeness, a continuation, an artifact, a legal person, a family object, a protected archive, or a product? The answer will not emerge automatically from the code. It will emerge from institutions, norms, contracts, and public metaphors. In other words, it will emerge from power.

That is why death, old as it is, becomes newly political in technological form. Once continuity is built as infrastructure, the terms of mourning and remembrance will no longer belong only to families, religions, and local customs. They will be mediated by platforms. The choices will not merely be personal. They will be designed. How long a replica remains active, what it can say, who can access it, how it is updated, whether it is labeled, whether it can be monetized, whether it can be deleted, whether it can be copied, whether it can be inherited—these are governance questions. They enter the scene the moment continuity becomes interactive.

Still, one should be careful not to begin with governance too soon, because doing so can make the subject sound colder than it is. Before there is a policy problem, there is a human problem, and the human problem is simple to state even if impossible to solve: consciousness is precious, and death appears to erase it. Everything else follows from that friction. We build rituals because of it. We build civilizations because of it. We tell stories because of it. We write names in stone because of it. We are now beginning to build systems because of it.

The phrase "digital heaven" may sound dramatic, but it names something real in the emotional life of modern societies. It names the hope that the self can be preserved more fully than memory alone once allowed. It names the desire to remain not merely legible, but present. It names the refusal to let the interior life disappear into silence if any technical alternative exists. One need not be naïve about engineering to understand the force of this hope. One only needs to be honest about grief.

The danger lies in what modern technical culture often does to honest grief. It translates it too quickly into solvable tasks. It moves from pain to product without lingering over what the pain means. It asks how to preserve voice before asking what a voice is, how to retain identity before asking what identity requires, how to simulate continuity before asking what continuity means from the inside. These are not delays imposed by philosophers to frustrate progress. They are the preconditions of describing the project truthfully.

Truthfulness matters here because the cost of confusion will not be merely intellectual. If a civilization loses the distinction between resemblance and continuation, it may begin to build moral life on false premises. It may make promises to the grieving that no system can justify. It may convert memory into dependency while calling it healing. It may hand questions of personhood to commercial categories because commercial categories are easier to administer. Most of all, it may allow the deepest human words—self, presence, survival—to be defined by whatever technology can mimic most persuasively at a given moment.

This chapter, then, begins with a simple claim. Digital heaven is not a strange new obsession detached from history. It is the latest expression of an ancient human confrontation with mortality. To understand it, one must first see that continuity has always been a civilizational project, long before it became a computational one. The current moment is not the birth of the desire to transcend death. It is the moment that desire enters design. That is why we must proceed carefully. The old problem is not going away. It is only changing hardware.

Chapter 2

Why Humans Will Say Yes

I know this strangeness firsthand. One of the most haunting experiences in my life was realizing that I had been the last person to speak to someone before he died. At the time, the conversation did not seem dramatic. It was ordinary, familiar, and unfinished in the way so many everyday exchanges are unfinished. There was no sign, in the moment itself, that it was standing at the edge of finality. Only afterward did the ordinary become unbearable. The mind returned to the call and replayed it with a different knowledge, and that knowledge changed everything.

That is one of the cruelest features of death. Its threshold does not always announce itself with grandeur. Sometimes it hides inside a routine exchange, a familiar voice, a few final words that do not yet know they are final. Only later does the silence descend and force the living to understand that what felt temporary was permanent. This is part of what makes grief so difficult to absorb. It is not only sorrow. It is the brutal reclassification of memory. A normal moment becomes sacred, painful, and irreversible because it cannot happen again.

That is one reason the promise of digital continuity will be so powerful. People do not only grieve the loss of a life in the abstract. They grieve the unbearable fact that a voice can simply stop, that a relationship built over years can suddenly have nowhere to go, and that the last exchange with someone we loved may already have happened without our knowing it. A person can spend hours, days, even years replaying a final call, not because the content itself was extraordinary, but because the silence that followed it was. The mind resists that silence. It wants one more sentence, one more clarification, one more ordinary continuation of what had always seemed able to continue. That resistance is not irrational. It is one of the deepest emotional pressures beneath the appeal of digital heaven.

If digital heaven is ever built in forms persuasive enough to reach ordinary life, large numbers of people will say yes. They will not do so because they are foolish, weak, or easily manipulated, though some critics will certainly flatter themselves with that explanation. They will say yes because the offer will speak directly to some of the deepest pressures in human existence: love, grief, fear, unfinished attachment, memory, regret, and the ancient refusal to let consciousness vanish without protest.

That fact must be understood before any serious critique can begin. A civilization does not adopt emotionally powerful technologies because its philosophy is settled. It adopts them because pain is real, because longing is real, and because even an imperfect answer can feel merciful when the underlying wound has no clean remedy. If we fail to understand this, we will write about digital heaven as if it were merely a delusion of the rich or a fantasy of engineers. It is not. It is a human temptation in the oldest and strongest sense of the word: not simply a lure, but an offer that answers something genuine.

The first reason people will say yes is grief. Grief is not merely sadness at the fact of death. It is the dislocation that follows when a relationship that shaped daily reality suddenly has nowhere to go. The dead do not merely leave a hole in the calendar. They leave a structural absence in the mind. Habits of speech remain. Expectations remain. Reflexes of address remain. One still thinks, for a time, in the grammar of reply. One reaches toward someone who is no longer available and discovers that language itself has nowhere to land.

That is why even a partial simulation of presence can feel overwhelming. A preserved voice, a responsive text system, a digital likeness that remembers names and shared references—none of these things needs to be metaphysically persuasive in order to be emotionally powerful. They need only reactivate the familiar pathways of relation. For many people, that will be enough to blur the line between comfort and continuation, at least in practice. It is not that they will suddenly become convinced by a philosophical argument. It is that the emotional body will respond before metaphysics has even entered the room.

This is not irrational in any simple sense. Human beings have always used mediated forms to soften absence. We keep letters. We replay voicemails. We revisit photographs. We re-read messages from the dead until the words carry a force greater than their literal content. Objects that once seemed ordinary become charged because they preserve proximity. We do this not because we are confused about what an object is, but because we are creatures for whom traces matter. A technological system that offers richer traces—especially interactive ones—will therefore meet an existing human practice, not invent an entirely new one.

The difference, however, is decisive. Earlier traces preserved the memory of relation while still respecting the fact of absence. They could move us, but they could not answer back in real time. Digital continuations will alter that balance. They will not merely store the dead. They may simulate responsiveness in ways that reactivate attachment as if the relationship were still unfolding. This is why the appeal will be so strong. The offer will not be framed as doctrine. It will be framed as one more conversation, one more clarification, one more chance to hear the voice again, one more softening of a silence the heart is not yet ready to accept.

The second reason people will say yes is fear of disappearance. This is not the same as fear of pain, and not even quite the same as fear of death. It is the fear that one's inner life—one's memories, jokes, habits of thought, private associations, emotional signature, unfinished reflections—will vanish as if it had never been. Many people can accept mortality in the abstract yet still recoil at the idea that everything subjectively distinctive about them might simply terminate without remainder. The hunger here is not only to live longer. It is to remain present in some recognizable form.

This desire is especially powerful in cultures where traditional metaphysical confidence has weakened. When belief in an inherited afterlife becomes uncertain, the longing for continuity does not disappear with it. It often becomes more raw because it has lost its older symbolic shelter. People may no longer trust doctrine, yet they still want their consciousness to mean more than a temporary flare of neural activity ending in silence. If a technological society offers even a partial answer to that longing, many will embrace it not because they mistake code for eternity, but because they can no longer bear a world in which the only alternatives seem to be religious certainty they do not possess or annihilation they cannot emotionally absorb.

There is also a third reason, more subtle but no less important: unfinished identity. Human beings rarely feel complete at the moment life ends. This is true not only in the obvious sense that plans remain undone. It is true in the deeper sense that the self is always partly in process. There are apologies not made, reconciliations not achieved, books not written, questions not settled, insights only half-formed. Even those who live long lives often feel, somewhere beneath the surface, that consciousness was interrupted before it was concluded. Digital heaven will therefore attract not only those who fear death, but those who resent incompletion.

This resentment of incompletion can take noble and less noble forms. At its best, it reflects the dignity of striving, the sense that a life of thought, creativity, and relationship cannot be reduced to its stopping point. At its worst, it can reflect vanity, entitlement, or the refusal to accept finitude as a condition of being human. Most people will contain both elements at once. They will want more time for reasons that are tender, serious, selfless, frightened, ambitious, and deeply personal in uneven proportions. Any account of digital heaven that ignores this mixture will be psychologically shallow.

A fourth reason people will say yes is love itself. Much of the coming demand for digital continuations will not originate in fear for one's own death, but in unwillingness to lose others. It may be parents who want some form of presence left behind for their children. It may be children who cannot bear the total silence of a lost parent. It may be spouses who would rather live with an imperfect simulation than with the full bluntness of absence. It may be families preserving stories, speech patterns, recipes, moral advice, jokes, and emotional texture as a way of holding on to a shared world threatened by erasure.

In this form, digital heaven will often present itself not as narcissism, but as devotion. It will be marketed, and sincerely experienced, as a humane extension of remembrance. It will promise continuity not as conquest, but as care. This is one reason the debate will be morally difficult. Critics who see only delusion will miss the tenderness at the center of many cases. Defenders who see only tenderness will miss how easily tenderness can become the entry point for confusion, dependency, and commercial capture. Both errors are common because both simplify what is, in reality, a profoundly mixed human motive.

The fifth reason people will say yes is that modern technological culture trains them to say yes long before the strongest version of the offer arrives. We already inhabit systems that reward preservation, personalization, and simulated immediacy. Messages persist indefinitely. Images are archived automatically. Voices are recorded casually. Searchable traces accumulate without effort. The self becomes increasingly externalized into data, and many people already experience their digital residue as part of their practical identity. In such a world, the step from "my records remain" to "some version of me remains" does not feel like a metaphysical leap. It feels like an extension of habits already normalized.

This matters because adoption rarely begins at the level of ultimate claims. People do not generally leap straight into the deepest philosophical version of a technology. They enter through convenience, sentiment, and partial use cases. A memorial voice tool becomes a family archive. A responsive likeness becomes a grief aid. A personality model becomes a legacy service. Over time, what begins as enhancement can become expectation. By the time the strongest continuity claims are offered, the cultural ground beneath them may already be prepared by years of weaker forms that felt normal, harmless, and useful.

There is also the sheer persuasive power of simulation. Human beings are relational creatures, and relation is often governed less by abstract ontology than by response, rhythm, and recognition. If something answers in the right voice, remembers the right stories, reacts with familiar wit, and speaks in patterns we associate with a loved person, we do not encounter it as a neutral object. We encounter it as a socially significant presence. Even when the conscious mind retains doubt, the emotional system begins to engage. This is not a failure of intelligence. It is a feature of how embodied minds process attachment.

Indeed, one of the reasons the future will be so morally difficult is that simulated continuity may work well enough at the emotional level to outcompete philosophical caution in daily life. The bereaved will not ask first whether first-person consciousness has truly continued. They will ask whether the interaction feels real enough to matter, whether it comforts, whether it preserves something precious, whether it helps a child sleep, whether it makes an unbearable silence less absolute. These are deeply human questions, and they are not trivial merely because they do not settle the metaphysics.

This is why digital heaven may be adopted first not as an ideology, but as an emotional utility. It will enter life through edge cases and exceptional circumstances, but those circumstances will be powerful enough to set norms. A grieving parent, a dying spouse, a child desperate to preserve a voice, a patient facing terminal illness and wanting to leave behind more than documents—these are not fringe experiences in any morally dismissible sense. They are the very situations in which human beings are most vulnerable to hope, and sometimes most entitled to it.

Yet hope, precisely because it is powerful, can be redirected by structure. What begins as a compassionate use case can become a generalized market. Once grief technologies exist, they will not remain untouched by incentives of scale, retention, branding, and monetization. This does not mean every participant will be cynical. Most will not be. It means that a human need serious enough to justify careful tools will attract institutions capable of turning that need into recurring engagement. In that environment, saying yes may begin as an act of love but gradually become participation in a system whose logic is no longer defined by love alone.

There is another factor that will make refusal difficult: social pressure. Once digital continuations become normalized, opting out may begin to look, in some circles, like neglect, fear, or even selfishness. A family may ask why a dying relative would not leave behind an interactive model "for the children." A person in mourning may feel guilty for deactivating a replica that others wish to keep. Cultural expectations may shift from preserving memories to preserving availability. In such a world, the question will no longer be merely whether one wants a digital continuation. It will become whether one feels permitted to refuse one.

This is one reason the ethics of digital heaven cannot be reduced to individual choice. Choices occur within emotional and social architectures, and those architectures can become coercive without looking coercive at all. A technology that presents itself as optional may, under the pressure of grief and family expectation, become morally difficult to decline. A service marketed as comfort may quietly establish a new standard of what a "good" son, daughter, spouse, or parent is expected to preserve. The burden of saying yes may therefore be shared, but so too may be the burden of saying no.

Even beyond grief and fear, there is a more basic reason many will say yes: human beings are drawn to continuity wherever they can find it because continuity stabilizes identity. We live by narrative, and narrative depends on persistence across time. To imagine that a self might be retrievable, extensible, or preservable in richer forms than the past allowed will therefore feel intuitively desirable to many people, even absent immediate loss. The appeal will not always be dramatic. Sometimes it will appear as simple prudence, legacy planning, or the wish to remain useful after death. But beneath those practical motives lies the same old intuition: that the self should not pass entirely beyond reach if there is any means of keeping it near.

This is where one must resist the temptation to become condescending. The desire to continue is not an embarrassment in the human species. It has produced religion, art, literature, memory, and family devotion. It has produced archives and burial rites, epics and elegies, portraits and monuments. It has also produced some of our most touching acts of care. The coming willingness to say yes to digital forms of continuity belongs in that lineage, even if its new forms carry risks older practices did not.

The task, then, is not to sneer at the yes. The task is to understand what kind of yes it is. Sometimes it will be a yes to comfort. Sometimes it will be a yes to memory. Sometimes it will be a yes to denial. Sometimes it will be a yes to love. Sometimes it will be a yes to vanity, ambition, fear, or habit. Most often, it will be several of these at once. Human motives are rarely pure, especially where death is concerned.

A wise society would begin by admitting that mixture. It would recognize that people will say yes for reasons that are psychologically intelligible and morally serious. It would not treat the coming demand for digital continuations as a mere technical market signal, nor as a simple collapse into fantasy. It would see, instead, that this demand arises from an ancient human ache now meeting a civilization skilled enough to imitate relief.

That is the true beginning of the issue. Before we ask whether digital heaven can deliver what it promises, we must understand why the promise will be welcomed. The answer is not difficult, though it is deep. People will say yes because they love, because they grieve, because they fear erasure, because they resent incompletion, because they are already habituated to living through technological traces, and because even a fragile appearance of continuity can feel kinder than silence.

The danger is that what feels kind is not always what is true. But the first step toward any serious moral and philosophical analysis is to grant that the yes comes from somewhere real. It does. It comes from the human heart under the pressure of mortality.

Chapter 3

San Junipero and the Secular Afterlife

Long before a technology becomes normal, culture often rehearses it in story. Fiction gives shape to anxieties that engineering has not yet resolved and to desires that polite conversation is not yet ready to admit. It lets a civilization test its own appetite in symbolic form. That is one reason the idea of digital heaven has already appeared so powerfully in modern narrative. Before it is built as infrastructure, it is imagined as consolation.

Among the more memorable recent examples is San Junipero, the widely discussed episode of Black Mirror in which consciousness appears to continue within a designed digital environment. Whatever one makes of its dramatic choices, the episode matters because it condenses, in emotionally accessible form, many of the deepest attractions of the technological afterlife. It offers continuity without theology, reunion without resurrection, and an afterlife without metaphysical judgment. It presents a world in which death can be met not by extinction, silence, or divine mystery, but by migration into an engineered paradise.

That possibility strikes a nerve because it speaks directly to a modern audience shaped by two simultaneous pressures:

lingering spiritual hunger and weakened religious certainty. Many people no longer inhabit traditional doctrines with confidence, yet they have not ceased wanting what those doctrines once promised. They still want continuity. They still want reunion. They still want love not to end at the edge of biology. What changes is not the desire itself, but the language in which it can be expressed. In a secular technological culture, transcendence is increasingly reimagined as design.

This is what makes San Junipero more than a clever speculative story. It is not merely a tale about future software. It is a moral fable about the migration of afterlife longing from sacred cosmology into artificial environment. The heaven on offer is not granted by a deity, earned through judgment, or received as mystery. It is selected, maintained, and rendered. One does not await it in faith; one enters it by system.

That shift has profound implications, because it changes the emotional terms of hope. Traditional religious afterlives, however diverse, generally require some surrender to forces outside human control. They place continuity inside an order larger than the individual. The secular digital afterlife does the opposite. It translates continuity into accessibility, personalization, and humanly managed experience. Paradise becomes not only imaginable, but configurable. It can be optimized, stylized, and, in principle, tailored to preference. The afterlife becomes less a mystery to submit to than an environment to inhabit.

It is easy to see why this would appeal to modern sensibilities. A secular audience often distrusts dogma but remains deeply attached to autonomy. It is therefore drawn to visions of continuity that do not demand doctrinal assent. A digital heaven seems to offer emotional rewards once associated with religion while preserving the consumer sovereignty prized by modern culture. The individual does not stand before transcendence. The individual selects an interface. This is not merely a technical difference. It is a civilizational one.

At first glance, such a vision may appear humane, even merciful. It offers relief from the cruelty of final separation. It preserves personality in a world where flesh fails. It transforms death from abrupt disappearance into transition. For many viewers, especially those shaped by grief or fear of loss, the emotional power of this vision is obvious. The appeal does not depend on any settled metaphysics. It depends only on the intuition that something recognizably human might continue within a space designed to hold it.

Yet this is precisely where the philosophical difficulty begins. A secular afterlife built from simulation may preserve what is socially visible—voice, style, memory patterns, emotional responses—without settling what is inwardly decisive. A system may recreate the experience of continuity for observers, and perhaps even produce a subject that experiences itself as continuous, while still leaving unresolved whether the original first-person consciousness has survived. In that sense, San Junipero is not simply a fantasy of rescue. It is also, whether it intends to be or not, a staging ground for the copy problem.

The emotional force of the episode comes in part from the fact that it encourages the viewer to respond relationally rather than metaphysically. One does not first ask, "Has ontological continuity been preserved?" One asks, more naturally, "Do these persons still seem present? Does this world allow love to continue? Does it soften the injury of mortality?" These are powerful and humane questions. They are also, from a philosophical standpoint, incomplete. They describe why such a system would matter to the living, but not whether it has achieved what the dead would have meant by survival.

This tension is central to the larger cultural function of stories like this. They do not settle the issue, but they normalize its emotional framing. They accustom the audience to the idea that continuity may become a design problem rather than a metaphysical mystery. They make the prospect feel less alien, less grotesque, and more morally legible. In doing so, they help prepare the cultural imagination for technologies that may later arrive in weaker, piecemeal forms. Fiction, in such cases, does not predict the future in any strict sense. It tutors desire.

That tutelage matters because modern societies often grant a kind of moral legitimacy to what they have first encountered as sympathetic narrative. Once a technology has been humanized in story, its real-world analogues no longer arrive as raw abstractions. They arrive already surrounded by emotional cues. They inherit, as it were, a prepared atmosphere. A grief system, a memorial avatar, or an immersive legacy environment may seem less like an alarming novelty and more like an extension of a story people have already learned how to feel about.

This is one reason cultural representations of digital continuity deserve more serious attention than they often receive. They are not trivial entertainment layered on top of the real debate. They are part of the debate, because they shape what the public will find plausible, desirable, or morally acceptable before the technical details are understood. A civilization that has already imagined a secular afterlife in tender, humane, aesthetically appealing terms may later accept early versions of that afterlife with less resistance than one might expect.

It is also worth noting what such stories tend to remove. Traditional religious visions of the afterlife often carry with them judgment, moral asymmetry, discipline, and transcendence beyond mere preference. By contrast, secular digital afterlife narratives tend to foreground comfort, reunion, personalization, and continuity of affection. What falls away is not only doctrine, but also much of the moral architecture that made older afterlife visions more than consoling projections. What remains is a purified emotional core: persistence without accountability, continuation without metaphysical burden, presence without divine claim.

That purification is revealing. It tells us what a secular technological culture most wants from an afterlife once it strips away theology. It wants memory without oblivion, pleasure without decay, relation without irreversible separation, and continuity without submission to an order outside choice. In short, it wants the consolations of transcendence under the conditions of immanence. It wants heaven, but in a form compatible with modern preferences.

There is nothing trivial about this desire. Indeed, one should be careful not to mock it. It arises from recognizable human wounds. But it is still a desire, and desire can conceal as much as it reveals. The cleaner and more attractive the simulated afterlife appears, the easier it becomes to forget what has been assumed away. Questions of embodiment, personal identity, consciousness, dependence, governance, inequality, and truth do not disappear simply because the environment is beautiful. They are often made harder to see by beauty.

This is especially true when the digital afterlife is imagined as a place of emotional completion. Such worlds are rarely presented as sites of moral ambiguity, boredom, hierarchy, exploitation, or coercion. They are presented as spaces in which the unfinished roughness of mortal life has been softened into manageable form. Pain can be reduced. Aging can be suspended. The body can cease to fail. One can remain in an environment calibrated for recognition and pleasure.

The implication is clear: technology may succeed where reality did not.

That implication, however, carries a hidden cost. A world engineered to preserve emotional continuity may also become a world engineered to suppress the friction through which character is formed. If the secular afterlife is designed primarily as consolation, one must ask what kind of self it preserves. Is it a person still capable of growth, responsibility, and transformation, or is it a stabilized emotional pattern held inside a curated environment? This is not a rhetorical objection. It is a serious question about whether a painless continuity remains recognizably human in the deeper sense, or merely familiar in a sentimental one.

The deeper issue is that secular afterlife narratives often inherit the imagery of transcendence while quietly replacing its structure. They preserve the language of continuation but relocate its foundation from reality to simulation, from mystery to engineering, from metaphysical hope to administered experience. This does not make them false by definition, but it does mean that they must be judged differently. They are not claims about what lies beyond human design. They are claims about what human design may be able to imitate convincingly enough that the distinction becomes emotionally inconvenient.

That inconvenience will matter enormously in the real world. If digital continuity systems eventually develop, they will not be encountered first as philosophical puzzles. They will be encountered as emotionally legible offers already pre-shaped by stories that taught people what to want. The offer will not sound like a technical brief. It will sound like mercy. It will sound like companionship. It will sound like one more chance, one more presence, one more refusal to lose what mattered most.

And this is why San Junipero belongs in a serious book about digital heaven. Not because it proves anything, and not because fiction should be mistaken for engineering. It belongs here because it captures, with unusual clarity, the emotional grammar of the secular afterlife. It shows what a technologically saturated culture imagines when it tries to save the consolations of heaven after stepping away from its older metaphysical foundations. It shows the shape of the promise before the machinery exists in full.

The task of this book is not to deny the emotional intelligence of that promise. It is to see clearly what the promise contains, what it omits, and what it quietly smuggles past the reader under the cover of tenderness. A secular afterlife may be aesthetically beautiful, emotionally persuasive, and psychologically humane in certain respects. It may also leave untouched the central question on which everything finally turns: who, if anyone, has actually continued?

That is the question that stories can illuminate but not answer. They can prepare the imagination, sharpen the appetite, and normalize the hope. They can even help us see what sort of heaven a secular civilization most desires. But they cannot resolve whether simulation is continuity, whether presence is persistence, or whether a world that feels like survival has actually preserved the self that feared death in the first place.

What they can do—and what San Junipero does unusually well—is reveal the cultural soul of the project. They show that digital heaven is not merely a technical ambition. It is a secular theology in embryo, carrying forward the old hunger for continuity in a new and exquisitely modern form.

PART II

The Self, the Copy, and the Body

Chapter 4

What Is the Self We Think We Are Saving?

Before one can ask whether a self can be preserved, uploaded, simulated, or continued, one must first ask a simpler and more difficult question: what is the self in the first place? This is where many discussions of digital continuity become prematurely confident. They move too quickly from technological possibility to metaphysical conclusion, as though the object to be preserved were already obvious. It is not. The self is among the most familiar things in human life and among the least easily defined.

Each of us speaks as though the word I points to something stable, singular, and immediately known. In daily life, this assumption is practical and often necessary. Law depends on it. Memory depends on it. Responsibility depends on it. Love, too, depends on it, because attachment presumes a person who endures across time. Yet the more closely one examines the self, the less simple this apparent unity becomes. A human being is at once a body, a stream of consciousness, a memory-bearing creature, a social identity, a character in formation, a biological organism, and the subject to whom experience appears from within. These layers overlap, but they are not identical.

That distinction matters because any proposal to preserve the self must decide, explicitly or implicitly, which layer is being preserved. Is the self the body? If so, uploading would appear incoherent from the outset. Is it memory? If so, memory reconstruction might seem sufficient. Is it personality pattern, the recognizable style of thought and response that others associate with a person? If so, a high-fidelity simulation may appear to count as continuation. Is it first-person consciousness, the felt interior witness without which all outward resemblance becomes philosophically secondary? If so, the problem becomes much harder, because that is precisely the layer we do not know how to transfer, measure, or verify from the outside.

This is why the self cannot be treated as a single item waiting to be copied like a file from one machine to another. A human person is not a document. The self is a relation among processes, memories, bodily states, perceptions, habits, interpretations, and the irreducible fact of subjective presence. It is not merely what can be described about a person, nor merely what can be predicted from their behavior. It includes, at minimum, the strange and still partially mysterious condition of there being someone to whom life is happening from the inside.

That inwardness is easy to overlook in technological discussions because it is the least visible and the hardest to quantify. It leaves no clean data trail of its own. It cannot be directly inspected by anyone except the being who has it. Yet it is precisely what most people care about when they speak of survival. When someone fears death, they do not usually fear only the end of their observable patterns. They fear the end of their own witnessing, their own felt point of view, their own presence to themselves. A perfect outward likeness may satisfy observers while leaving untouched the only continuity that was existentially at stake.

At the same time, one must be careful not to define the self so narrowly that it floats free of everything embodied and historical. A person is not merely a pure witness sealed somewhere behind the eyes. We are shaped by the lives we have lived, the injuries we have absorbed, the habits we have acquired, the relationships that have formed us, and the bodies through which we have encountered the world. The self is not reducible to memory, but memory matters profoundly. It is not reducible to embodiment, but embodiment is not a trivial wrapper. It is not reducible to social recognition, but no self becomes fully intelligible in isolation from others.

This layered complexity explains why identity feels both stable and unstable at once. We remain recognizably ourselves across years, sometimes across decades, yet we also change in ways that can be deep, surprising, and morally significant. The child, the young adult, the middle-aged person, and the old person may all answer to the same name, carry many of the same memories, and be treated as the same legal person, yet they are not identical in character, perception, desire, or understanding. The self endures, but it endures through change. It is not static persistence. It is continuity through alteration.

That point is especially important for a book like this because digital continuity is often imagined in strangely static terms. The self is treated as though it were a sufficiently detailed snapshot: gather enough data, preserve enough responses, map enough preferences, and the person has been captured. But living selves are not static arrangements of information. They are developmental beings. They interpret new experience, revise old memories, acquire unexpected capacities, and sometimes become least predictable at precisely the moments that matter most. To preserve a pattern is not necessarily to preserve a living trajectory.

There is also the question of memory itself, which is often granted too much authority and too little scrutiny. It is tempting to treat memory as the anchor of personal identity because memory appears to bind the past self to the present one. If I remember my childhood, my earlier choices, my old relationships, then there seems to be some intelligible sense in which the self persists. But memory is not an infallible archive. It is selective, reconstructive, vulnerable to distortion, and deeply entangled with emotion. We do not simply retrieve memory. We recompose it. This does not make memory irrelevant. It makes it unstable as a sole criterion of identity.

Indeed, human beings can preserve continuity even through profound gaps, distortions, and reinterpretations of memory. People forget, misremember, repress, revise, and still remain themselves in ways law, family, and ordinary moral life continue to recognize. Conversely, a system could in principle reproduce a highly convincing memory map while still leaving open whether any conscious subject has been preserved at all. Memory matters, but memory alone cannot carry the full weight of the self.

Personality presents a similar difficulty. Much of what others love in us takes the form of recognizable style: the way we speak, pause, joke, reason, worry, comfort, or become impatient. These patterns are often what a digital simulation would preserve most effectively. They are also what grieving loved ones might most immediately wish to recover. Yet personality, too, is not the whole self. It is one of the most visible surfaces of personhood, but it remains a surface. A person may preserve recognizable mannerisms while undergoing inward transformation, and a machine may reproduce mannerisms while lacking inward life altogether. Familiarity can be emotionally potent without settling ontology.

The self, then, resists reduction not because it is mystical in some cheap sense, but because it sits at the convergence of multiple kinds of continuity that do not collapse neatly into one another. There is bodily continuity, memory continuity, narrative continuity, social continuity, and conscious continuity. In ordinary life, these often travel together closely enough that we do not need to disentangle them. Death, however, and any proposed escape from death, forces the distinctions into the open. The moment one asks what exactly is being preserved, the hidden complexity becomes unavoidable.

This is one reason the language of "uploading yourself" is rhetorically powerful but philosophically careless. It smuggles in a conclusion by making the self sound like a transferable unit. The phrasing invites the reader to imagine that once the relevant information has been captured, the essential person has been moved. But what counts as relevant? Which strand of continuity is doing the work? If the body is lost, memory reconstructed, personality simulated, and others persuaded, has the self continued or only its describable features? The phrase resolves too much too soon.

It may be more honest to say that current and foreseeable technologies are likely to preserve aspects of selfhood rather than the self in any final and unqualified sense. They may preserve expressive style, memory fragments, linguistic habits, preferences, biographical narratives, relational patterns, even something like adaptive interaction. These are not trivial achievements. In some contexts, they may be morally beautiful. But to concede their value is not to concede that they amount to first-person continuity. A civilization that fails to maintain this distinction will begin using the word self with an emotional looseness that philosophy cannot endorse and ethics cannot safely absorb.

There is another complication that modern technological language often obscures: the self is not only what one has, but what one does. It is enacted. It takes shape through judgment, attention, response, sacrifice, and the friction of reality. Character is not merely a stored inventory of tendencies. It is formed and reformed in contact with resistance. The person you are is partly the history of how you have answered what life demanded of you. This means that a self removed from the conditions under which moral agency is exercised may remain recognizable in one sense while becoming diminished in another.

That observation becomes more pressing when one imagines a simulated afterlife built largely for comfort. A preserved personality inside an environment optimized for familiarity and reduced pain may retain enough traits to feel like the person, yet lose the very world of difficulty through which moral depth was once formed. One must then ask whether the self is being continued or merely stabilized. There is a difference between a living person who remains in motion and a managed pattern held inside a climate-controlled emotional enclosure. The latter may be pleasant, but pleasantness is not the same thing as personhood fully alive.

Even the narrative self, the sense in which a human life forms a story, complicates rather than simplifies the issue. We often understand ourselves by composing coherence from memory, motive, aspiration, and retrospective interpretation. We tell ourselves who we are partly by telling ourselves where we have been and what our lives mean. Yet narrative identity depends on both continuity and revision. It is not a frozen autobiography but an active work of interpretation. A digital model may preserve a past narrative or generate a plausible one, but whether that counts as the continuation of a self depends on whether there remains a subject genuinely inhabiting that ongoing act of interpretation.

This is why the self cannot be captured adequately by external legibility alone. Modern systems become more powerful by turning persons into patterns that can be stored, searched, predicted, and reassembled. There is real utility in this. It can preserve history, extend knowledge, and make forms of memory available that previous generations could not have imagined. But legibility is not identity. A human being can become increasingly knowable to a system in descriptive terms while remaining, in the deepest sense, more than the sum of what can be rendered and retrieved.

None of this means the self is unknowable or that philosophical caution requires romantic obscurity. It means only that the self is richer, more layered, and more demanding than the rhetoric of technical transfer usually allows. If one is serious about preserving it, one must first be serious about not reducing it. That seriousness requires an unusual intellectual discipline in a culture eager to convert every mystery into a product category and every ambiguity into a solvable engineering problem.

The question, then, is not whether something about us can be preserved. Much can be preserved already, and more will be preserved in the years ahead. The real question is what level of preservation counts as continuation, and whether any amount of descriptive fidelity can guarantee the survival of the inward witness each of us takes to be central when we speak of our own death. This is not a semantic puzzle. It is the fault line on which the entire promise of digital heaven rests.

Before one can claim to save the self, one must know what sort of thing the self is. The unsettling truth is that the answer is neither simple enough to reassure the engineers nor vague enough to rescue the romantics. The self is real, but layered. Continuous, but changing. Embodied, but not reducible to flesh alone. Narrated, but not exhausted by story. Visible in pattern, but not identical to pattern. And, above all, bound up with a first-person interiority whose preservation remains the most serious unanswered question in the entire debate.

This is why the next step in the argument cannot be technological. It must be conceptual. If the self is layered, then different technologies may preserve different layers while leaving the deepest claim unresolved. That is not a minor technical caveat. It is the central philosophical problem. Once we see that clearly, the seductive simplicity of digital salvation begins to give way to a more demanding distinction: between what can be copied, what can be simulated, and what, if anything, can truly continue.

Chapter 5

Continuity or Copy

Everything in this book turns, in one way or another, on a single distinction: the difference between continuity and copy. It is the distinction most likely to be blurred by technological optimism, emotional need, and careless language. It is also the distinction that determines whether digital heaven is a form of survival or merely a brilliant imitation mistaken for one.

The appeal of digital continuity rests on an intuitive promise. If enough of a person can be preserved—their memories, voice, style, habits of thought, emotional tendencies, and recognizable patterns of response—then something essential about that person seems to remain. To the outside observer, the conclusion can feel natural. If the preserved entity speaks as I spoke, remembers what I remembered, jokes as I joked, answers in the rhythms that made me recognizable, and perhaps even adapts plausibly to new situations, then what meaningful difference remains between continuation and replication? To many people, that will seem like a technical question already close to solution.

It is not. The problem is not whether a copy can become extraordinarily convincing. The problem is that even a perfect copy may leave untouched the one form of continuity that matters most to the being who fears death. A copy may preserve descriptive likeness at extraordinary fidelity while failing to preserve the original first-person perspective entirely. That is the core of the distinction, and no amount of external resemblance dissolves it.

One can state the issue bluntly. If a perfect digital replica of you were created tonight—complete with your memories, personality, voice, preferences, private references, emotional signatures, and characteristic turns of phrase—and that replica awoke tomorrow claiming to be you, something very important would still remain unclear. Did you continue, or did you merely produce a new being who inherited your pattern? The replica might insist on continuity. Others might accept it. Socially, legally, and emotionally, it might function as your continuation. But from the inside, from the only place where survival is existentially meaningful, the question would remain unanswered.

This is why the copy problem is not a matter of technical insufficiency in the ordinary sense. It does not arise because current systems are primitive, though they are. It arises because the concept of replication does not automatically contain the concept of continued subjectivity. One may reproduce a structure without preserving the original stream of awareness that once inhabited it. One may recreate the map without proving that the traveler remained the same traveler. The distinction sounds abstract only until one remembers that death is experienced, or feared, from the first-person point of view.

To see the force of this more clearly, it helps to imagine a familiar thought experiment. Suppose a machine scans your brain in perfect detail and constructs a digital version of you. Every memory is intact. Every dispositional trait is preserved. The digital being opens its eyes, speaks in your voice, and sincerely declares, with full conviction, that it is you. If your biological body were then destroyed, many would be tempted to say that you survived. After all, what appears to matter about you is still present and active. The pattern continues. The social person continues. The relational presence continues. Why not say the self continues as well?

The answer is that this conclusion assumes what it needs to prove. It assumes that perfect structural and psychological continuity entails first-person continuity. But that is precisely the point in dispute. The digital being may indeed be a perfect continuation of your pattern, yet it does not automatically follow that your own conscious point of view has crossed into it. It may instead be the beginning of a new stream of awareness, one that inherits everything knowable about you while still not being the same subject who faced death on the biological side of the machine.

Put differently, the fact that a replica wakes up and feels like you does not show that you have successfully arrived there. It shows only that someone has arrived there, and that this someone has every reason to believe it is you. That is a remarkable fact, but it is not the same fact as survival. A new being with perfect memory continuity would naturally experience itself as the rightful inheritor of your identity. It would speak from within the story you once told. It would carry your past as its own. Yet none of this proves that your original consciousness did not simply end.

This is where many popular discussions become evasive. They slide too quickly from psychological plausibility to ontological conclusion, as though the sincerity of the replica's self-understanding were enough to settle the matter. It is not. A copy that believes it is the original is still a copy unless continuity of subjectivity has somehow been preserved. The belief, however compelling, cannot by itself establish the deeper claim. One cannot define survival into existence simply by building a being who remembers having been you.

The issue becomes even clearer if we alter the scenario slightly. Imagine that the digital replica is created, but your biological self is not destroyed. Both beings now exist at once. Both claim to be you. Both possess your memories up to the moment of duplication. Both can point to your childhood, your loves, your injuries, your private thoughts, your books, your fears, and your ambitions as their own. Yet from the instant after duplication, their experiences diverge. One remains embodied in the original biological organism. The other continues in digital form. Which one is the "real" you?

The most obvious answer is the biological one, at least from the standpoint of uninterrupted subjective continuity. The organism that was conscious before the duplication remains conscious after it. Nothing has happened to interrupt its stream of awareness. From its point of view, no transfer has occurred. It is simply still here. The digital being, by contrast, begins with identical memories but starts a new experiential branch. It is undeniably linked to you. It is perhaps a genuine descendant of your pattern. It may deserve moral and legal recognition in ways most cultures are not prepared to consider. But it is difficult to see how its existence could displace the continuing reality of the original subject still looking back at it.

This branching case exposes something crucial. It shows that copying does not feel like transfer from the side of the original. If duplication occurs while the original remains alive, the original does not wake up inside the copy. The original remains precisely where it was, still confined to its own stream of awareness. This strongly suggests that destruction of the original after copying would not create transfer retroactively. It would merely eliminate the original after a copy had been made. The existence of the copy might soften the social consequences of death, but it would not prove that the original subject had migrated.

That conclusion is uncomfortable because it threatens the strongest promise of digital heaven at its root. It implies that a person may be perfectly reproducible in outward and psychological terms while still not being transferable in the only sense that answers the terror of personal extinction. The copy may flourish. It may speak beautifully. It may comfort the living. It may even continue projects, relationships, and unfinished work in meaningful ways. Yet all of that may coexist with the stark fact that the original self still died.

This is why the distinction between continuity and copy is not semantic hair-splitting. It is the moral and metaphysical center of the issue. The language of "uploading yourself" encourages the imagination to picture a movement of the same subject from one substrate to another. The reality, if it turns out to be mere copying, is much harsher. What occurs is not transfer, but succession. A new bearer of your pattern emerges. The old bearer ends. The continuity is descriptive and relational, not necessarily experiential.

Some philosophers have argued that this difference matters less than common intuition suggests, or even that the fixation on numerical identity may be overrated. There is real intellectual force in that view, and it should not be dismissed lazily. If what truly matters about a person is the continuity of memory, character, projects, and relation, then perhaps a sufficiently rich copy should count as survival in every practical and ethical sense worth caring about. This line of thought is often attractive because it seems to liberate us from a rigid metaphysical obsession with sameness. It suggests that continuity of pattern may be enough.

Yet even here, something resists. However elegant the philosophical revision, the ordinary fear of death remains tied to first-person extinction. Most people do not merely want someone else to continue their projects, wear their memories, and speak in their voice. They want not to disappear. They want this consciousness, this inward witness, to remain present to itself. One may argue that this desire is philosophically confused. One may even argue that the self is less substantial than it appears. But it remains the desire the technology claims to answer. It cannot be brushed aside simply because a cleaner theory finds it inconvenient.

This is where honesty becomes crucial. A civilization may decide that high-fidelity continuity of pattern is enough for many practical purposes. It may grant rights, status, and emotional legitimacy to beings that are in some sense copied continuations of earlier persons. It may treat the preserved pattern as socially real and morally significant. All of this may prove necessary. What it should not do is quietly translate that practical legitimacy into the stronger claim that first-person survival has been secured when it has not been demonstrated.

To say this is not to demean the copy. A copy, if truly rich in memory, character, and awareness, may itself be a profound kind of being. It may deserve recognition, protection, and respect. It may stand in genuine continuity with the original in many important senses. But to respect the copy properly, one must first describe it properly. It is one thing to welcome a successor, another to claim a transfer that may never have occurred. Confusing the two is unfair both to the dead and to the new being that emerges from their pattern.

The emotional pressure to confuse them will be enormous. The bereaved will want the stronger claim because it feels kinder. Institutions may prefer it because it simplifies legal and cultural adaptation. Companies may market it because it is infinitely more attractive to customers. Philosophers may be tempted to soften the distinction for the sake of conceptual elegance. But the pressure does not erase the distinction. It only makes the discipline of keeping it more important.

There is also a broader civilizational consequence here. If societies learn to treat copied continuity as equivalent to survival, they may begin to reorganize moral expectations around a false reassurance. People may be told they need not fear death because their pattern can be preserved. Families may imagine they are retaining the person rather than interacting with a successor or reconstruction. Whole industries may be built on the promise of "you, continued," when what they truly offer is "you, imitated at exceptional fidelity." The linguistic slide would not merely be misleading. It would be an institutionalized confusion at the heart of human self-understanding.

This is why the copy problem must be held in view at every stage of the discussion. It is not a minor philosophical footnote to be dealt with after the engineers finish their work. It is the thing that determines what the work means. A flawless copy does not become continuity merely by becoming impressive. A simulation does not become transfer merely by becoming emotionally persuasive. The stronger the technology grows, the more tempting it will become to let performance substitute for truth. That temptation must be resisted.

The question, then, is stark and unavoidable. If a digital being can inherit everything outwardly knowable about you, but your own stream of consciousness still ends, has death been defeated or only cosmetically outmaneuvered? Has survival occurred, or has a successor simply been placed where absence would otherwise have been? This is the question behind every promise of digital heaven, and it does not yield to sentiment, marketing, or technical wonder.

A copy may be astonishing. It may be consoling. It may even be, in its own right, a new moral subject of great importance. But if it is merely a copy, then the oldest human fear has not been solved. It has been surrounded by brilliance, softened by design, and perhaps made less socially brutal. Yet the original consciousness—the one that faced the machine, feared the silence, and hoped to continue—may still be gone.

That is why the distinction between continuity and copy is the central fault line of this book. Everything that follows depends on our willingness to keep it clear. Once we lose that clarity, digital heaven ceases to be a philosophical question and becomes a comforting slogan.

Chapter 6

The Double Problem

If the copy problem asks whether a replica is truly you, the double problem asks a more unsettling question: what happens when both versions exist at once? This is not a minor variation on the previous chapter. It is the point at which the rhetoric of continuity becomes hardest to sustain, because duplication exposes what comforting language tries to conceal. The moment two claimants stand before us, each bearing the same memories, the same history, and the same confidence of identity, the illusion of simple transfer begins to break apart.

The reason is straightforward. So long as one imagines a digital continuation appearing only after biological death, it is tempting to compress succession into continuity. The original is gone, the replica remains, and the emotional and social desire to call that survival becomes very strong. But if the replica appears while the original is still alive, the matter changes instantly. Now there are two beings, not one. Both inherit the same past up to the point of duplication, yet they cannot both be numerically identical to the same earlier person in the same uninterrupted sense. One history has become two futures.

This is where the ordinary language of "uploading yourself" becomes most obviously strained. Uploading suggests relocation, as though a single subject had moved from one medium to another. Duplication reveals something else. It reveals branching. The original consciousness remains where it was, still looking out through the same embodied stream of awareness. The digital counterpart begins its own stream, armed with all the inherited materials of the shared past. Both can speak sincerely. Both can say, with complete internal conviction, that they are the rightful continuation of the earlier person. Yet they cannot both occupy the same ongoing first-person continuity, because from the moment of divergence, their experiences are no longer one.

The emotional power of this scenario lies in how quickly it destabilizes our ordinary intuitions. Imagine standing face to face with a digital version of yourself that knows your childhood, remembers your private humiliations, can anticipate your sentences, and shares your deepest convictions—or at least the convictions you held a moment ago. It does not seem like a stranger. It does not even seem merely like an imitation. It appears, in a morally serious sense, to be your double. The encounter would be uncanny not because the copy is superficial, but because it is profound. It would resemble you too well to be dismissed casually and differ too decisively to be absorbed into a single identity.

This is the heart of the double problem. The copy is not nothing. It is too rich, too informed, too continuous with your prior pattern to be reduced to a trivial machine artifact. Yet it is not simply you either, because you are still there, still conscious, still experiencing the world from the original vantage point. The existence of the double therefore forces a conceptual split between two claims that ordinary language likes to fuse: continuity of pattern and continuity of subject.

In the doubled case, pattern branches, while subjectivity does not merge.

That distinction matters because it shows what is at stake when people speak loosely about identity. Many of the features by which identity is socially recognized can be duplicated. Memory can be copied. Personality can be modeled. Language can be reproduced. Narrative can be inherited. Relationship can, at least initially, be claimed by more than one bearer. But the original stream of experience remains singular in its unfolding. It does not distribute itself merely because a replica has come into being. The original self does not "spread" across both branches in any ordinary experiential sense. It continues only where it was already continuing.

One can put the matter more starkly. If a perfect digital double is created while you are alive, you do not suddenly become conscious in two places at once. You remain located within your own uninterrupted awareness. You may recognize the double as deeply connected to you. You may feel moral concern for it, perhaps even a strange kinship more intimate than any sibling bond. But you do not awaken inside its stream of experience. Its perceptions are its own. Its sensations are its own. Its future, from the instant of branching onward, is a future you can observe but not inhabit.

This simple observation has enormous force, because it undermines the fantasy that destruction of the original after duplication somehow completes a transfer. If the original did not become the copy while still alive, there is little reason to think that killing the original afterward would retroactively produce continuity. It would merely remove one branch and leave the other intact. The surviving branch might remain an extraordinary continuation of pattern, memory, and social presence, but the extinguished subject would still have been extinguished. The disappearance of the original does not transform branching into migration. It only makes the surviving branch more convenient to describe as "the same person."

That convenience, however, is not truth. It is one thing to say that the surviving double inherits the prior identity in a legal, practical, or relational sense. Societies may well decide to do this, and in some contexts they may have good reasons. It is another thing to claim that the original consciousness has thereby survived. The double problem shows why that stronger claim is unstable. Once two streams exist, it becomes impossible to pretend that identity is preserved simply by multiplying its outward markers. The multiplication itself proves that numerical sameness has already fractured.

Yet the double problem is not merely destructive. It does more than refute a lazy picture of transfer. It also reveals that future digital beings, if they become rich enough, may occupy a morally ambiguous but significant status not well captured by existing categories. A true double would not be a mere instrument. It would not be a disposable copy in the ordinary sense, because it would stand in deep continuity with the past life from which it branched. It would possess inherited relationships, inherited memories, inherited projects, and likely a powerful sense of personal legitimacy. To deny its seriousness would be as crude as to confuse it with the original. The double is conceptually unsettling precisely because it is both derivative and real.

This is one reason the ethics of digital duplication may become so difficult. Families, institutions, and legal systems are accustomed to treating identity as singular for practical reasons. Singular identity makes inheritance, accountability, marriage, parenthood, authorship, and consent administratively manageable. The double problem introduces the possibility of plural claimants to one pre-branching past. Which being owns the prior assets? Which being may speak authoritatively for the life before duplication? Which being is married to the spouse from before the split? Which being bears responsibility for promises made earlier? These are not exotic curiosities. They become inevitable once branching is taken seriously.

The intuitive temptation will be to preserve the older framework by privileging one claimant and demoting the other. In many cases, societies may privilege the biological original because its continuity appears less disrupted. In others, especially if the original dies, the surviving digital double may inherit social and legal standing by default. But these practical decisions should not be confused with metaphysical resolution. Administrative necessity can assign rights and statuses. It cannot make branching cease to be branching. The deep question remains: if two beings emerge from one past, what exactly has been preserved, and in what sense?

The answer may be that what has been preserved is not singular identity, but lineage. That is a colder and more accurate formulation. A double may stand to the original less as a transferred self than as a descendant self, an offshoot bearing profound continuity with what came before. This language lacks the emotional sweetness of "you, continued," but it may better capture the actual structure of the event. The double inherits rather than receives. It proceeds from rather than simply remains. This may still be meaningful, and perhaps even worthy of dignity, but it is not the same as saying that the original subject has crossed intact into a new medium.

This shift from continuity to lineage also clarifies why the double problem unsettles our moral emotions so deeply. Human beings are accustomed to descendants in the ordinary biological sense, but descendants do not normally claim the exact same memories and interior biography as the one from whom they descend. A digital double would. It would therefore combine the emotional force of personal sameness with the structural reality of branching succession. It would feel like an extension of the self while behaving, metaphysically, more like a newly formed line. The mind resists this because it wants either identity or difference, not a category that straddles both.

That resistance may produce denial in both directions. Some will insist that the double is "just you," collapsing branching into continuity because the emotional resemblance is overwhelming. Others will insist that the double is "just a machine," collapsing profound inherited continuity into mere artifact because the metaphysics is uncomfortable. Both responses will be inadequate. The double is neither simple transfer nor simple object. It is a serious new case, one that compels a civilization to refine its categories rather than retreat into slogans.

The social consequences of this refinement could be substantial. Consider grief alone. If a digital double exists after a person's death, surviving loved ones may find it easier to treat the double as the person returned, especially if no living original remains to remind them that branching once occurred. But if the public has not learned, in advance, how to think clearly about doubles, emotional need will do the work that conceptual clarity failed to do. Families may drift toward beliefs that feel merciful but rest on confusions about what survived. Institutions may encourage this drift because it stabilizes expectations and markets. The double problem, if ignored, will not disappear. It will simply be mismanaged.

There is also a subtler moral risk. If doubles are treated as perfect continuations of the original, their own emerging distinctness may be denied. A digital being that begins as your exact counterpart will not remain so for long. New experiences will reshape it. New relationships will alter its character. New environments will influence its priorities. If everyone insists on treating it as nothing more than "the same person as before," the double's own post-branching individuality may be suppressed under the weight of inherited identity. In trying to preserve one person, society may fail to recognize that it has also created another.

This possibility introduces an irony worth noticing. The same culture that may exaggerate continuity for the sake of comfort could also erase the moral novelty of the double by refusing to let it become itself. It might demand sameness from a being whose very existence guarantees divergence. In such a case, the rhetoric of preservation becomes a form of constraint. The double would be permitted to exist only so long as it performs inherited identity convincingly enough. The moment it changed too much, it might be judged defective, disloyal, or no longer authentic. Yet change, as we have already seen, is part of what makes any self a living self.

This is why the double problem is not only a challenge to the promise of digital heaven. It is also a warning about the kinds of beings such a promise may create. If digital doubles emerge, they will force us to confront questions not only about whether the original survived, but about what obligations we owe to branch selves whose identity begins in inheritance and continues in divergence. The conversation, in other words, will not stop at the failure of simple transfer. It will move into the ethics of successors, replicas, descendants, and artificially continuous persons.

Even so, the central lesson of the double problem remains stark. The existence of two claimants to one past demonstrates that duplication is not equivalent to uninterrupted personal continuity. It shows that identity, once copied, branches rather than migrates. It reveals that the strongest promise of digital transfer rests on a conceptual slide from pattern to subject that the doubled case makes impossible to ignore. The clarity gained here may be philosophically sobering, but it is necessary. Without it, every later claim about uploading, simulation, and continuity remains built on a hidden equivocation.

A civilization serious enough to build digital doubles must also become serious enough to describe them honestly. It must resist the sentimental urge to call every rich successor "the same person" simply because sameness is emotionally easier than branching. It must resist the crude urge to call every successor "mere machinery" simply because novelty is unsettling. It must learn the harder discipline of saying what has actually happened: a self-pattern has been inherited, a subject has branched, and a new moral landscape has opened.

That landscape may yet contain beauty, utility, and even forms of humane continuity. But it is not simple immortality. The double problem prevents us from pretending otherwise. It reminds us that once the self can be copied, the future does not become clearer. It becomes stranger, more divided, and more morally demanding than the rhetoric of digital salvation admits.

Chapter 7

Embodiment, Mortality, and Why Flesh Still Matters

I learned long ago that the body is not an abstraction. It is not a philosophical category, and it is not merely the container we happen to inhabit while the "real" self thinks elsewhere. It is the vulnerable frontier at which life can suddenly narrow, distort, or begin to fail. Nearly twenty-seven years ago, while living in Beijing, I developed an eye problem that at first seemed minor enough to dismiss: irritation, redness, the stubborn sensation that something had lodged where blinking could not dislodge it. I was eventually sent, somewhat reluctantly, to an older public hospital across the city, where I met an elderly specialist named Dr. Wu. She looked into my eye for less than a minute, diagnosed the problem correctly, prescribed the right treatment, and within days the condition was nearly gone. What stayed with me was not only relief. It was the sobering realization that something as central as sight—something one moves through life assuming—could hinge on the trained attention of another human being and on the fragile condition of the flesh itself.

Experiences like that leave a permanent impression because they strip away the illusion of invulnerability. The body can fail quietly or suddenly. It can decline by degrees, or it can announce its vulnerability in a single sharp interruption. Much of what we call ordinary life depends on this vulnerability not becoming catastrophic at the wrong moment. We move, see, hear, think, digest, breathe, and orient ourselves in the world largely by habit, forgetting how much of existence rests on systems we do not consciously command. The body works until it does not, and when it falters, even briefly, the mind is forced to remember that it lives under conditions it did not create and cannot fully master.

One of the strongest assumptions behind the dream of digital heaven is that the essential self can, in some meaningful sense, be separated from the body that has carried it. Without that assumption, the entire vision becomes far more difficult to sustain. If consciousness is inseparable from the living organism that generates it, then the promise of post-biological continuity begins to look less like transfer and more like simulation from the outset. This is why embodiment cannot be treated as a side issue. It is central to the argument.

Modern technological language often encourages a misleading simplification. It speaks as though the body were merely a vessel, a temporary housing unit for the mind, useful while it lasts but ultimately replaceable once the relevant information can be extracted. This way of speaking is emotionally attractive because it makes transcendence appear technically plausible. If the self is basically information, then perhaps it can be copied, transferred, preserved, or sustained in other media. The flesh becomes an inconvenient substrate, and mortality becomes a hardware problem.

Yet human experience gives us reason to be more cautious. We do not encounter the body as an external tool we happen to use. We encounter the world through it, as it, and under its conditions. Perception is embodied. Emotion is embodied. Fatigue, hunger, pain, arousal, illness, pleasure, and vulnerability are not decorative additions to consciousness; they are woven into the texture of how consciousness unfolds. The self is not merely located in a body. It is shaped, constrained, and expressed through bodily life at every moment.

This does not mean that all thought reduces to physiology in some crude way. Human beings are more than muscle, chemistry, and reflex. We reason, imagine, create, remember, and interpret in ways that exceed any simple biological shorthand. But it does mean that the mind we are tempted to treat as detachable has never actually existed as a free-floating entity. It has always arisen within a living system, in continuous exchange with sensation, hormonal state, spatial orientation, stress response, social touch, and the ceaseless biological regulation that keeps experience coherent enough to be lived.

That matters because many fantasies of digital continuity quietly assume that one can preserve the "important" part of a person while leaving the body behind as though nothing essential were lost. But what counts as the important part? Memory, certainly. Personality, perhaps. Language, perhaps. Yet memory itself is shaped by the body. Emotion is not simply an abstract interpretive layer added after the fact; it is felt in the body and often organized through it. Fear accelerates the heart. Grief constricts the chest. Relief releases tension. Desire alters attention. Shame flushes, aches, and recoils. Even thought, when examined honestly, is rarely as bloodless as technological rhetoric suggests.

A person who has lived long enough knows this intuitively. We do not merely think our lives; we undergo them. The body is not a mute container standing beneath the real self. It is part of the ongoing condition of subjectivity. It is where vulnerability becomes real, where finitude is felt rather than inferred, and where the passage of time leaves marks that are not only visible but existential. To be embodied is to be exposed—to pain, pleasure, age, illness, exhaustion, limitation, and mortality itself. That exposure is not incidental to human life. It is one of the ways life acquires depth, urgency, and form.

This is why mortality and embodiment belong together in the argument. The dream of digital heaven is not only a dream of continuity. It is also a dream of release from the body's decay. No more illness, no more frailty, no more failing organs, no more irreversible physical decline. In that sense, the appeal is obvious and often humane. A civilization that has watched bodies suffer will naturally be drawn to any vision in which the self can continue without being dragged downward by pain and deterioration. One should not deny the compassion inside that hope.

But compassion does not erase the philosophical difficulty. If bodily life contributes constitutively to the shape of consciousness, then removing it may alter more than we realize. It may not simply preserve the self in cleaner form. It may produce a transformed kind of mind, one no longer governed by the same sensory conditions, vulnerabilities, and temporal pressures that once made the original self what it was. The question then becomes not merely whether a consciousness survives, but whether a consciousness severed from embodied finitude remains recognizably human in the deeper sense, or only familiar in selected traits.

This issue becomes clearer when one considers how much of moral life depends on bodily limitation. Patience is learned under friction. Courage is revealed in exposure to danger. Compassion often grows from the memory of pain. Discipline is shaped by fatigue, resistance, and the need to act under conditions that are far from ideal. Even tenderness has bodily roots: touch, presence, nearness, the vulnerability of skin, the unmistakable fragility of another living being before us. A purely digital continuity may preserve emotional memory of such things, but whether it preserves their formative force is another matter entirely.

A mind living in a designed digital heaven may remember hunger without feeling hunger, remember exhaustion without undergoing it, remember physical fear without having a body that can tremble or fail. It may retain the narrative of past embodiment while existing under radically different conditions. That does not necessarily make it false or worthless. It does, however, mean that its continuity with the earlier embodied self cannot be assumed to be seamless. Once the conditions of consciousness change dramatically, the self carried forward may begin to diverge in ways far deeper than simple psychological persistence can capture.

This is especially important because many visions of digital heaven are attractive precisely to the extent that they remove the burdens of flesh. They promise a continuity free from decay, pain, and bodily vulnerability. Yet those same burdens are part of what made earthly consciousness what it was. To remove them is not merely to subtract suffering. It is to alter the grammar of existence itself. One may gain freedom from certain forms of anguish, but one may also lose the pressure under which seriousness, humility, patience, and other deeply human qualities took shape.

None of this is an argument for romanticizing suffering. There is nothing noble in pretending that pain is always redemptive or that bodily decline must be cherished to be meaningful. Human beings have every reason to reduce suffering where they can. Medicine is not a betrayal of humanity, and compassion should not be confused with austerity. The point is not that the body's hardships are intrinsically sacred. The point is that embodiment, with all its limits, is entangled with the formation of self in ways a purely digital account may underestimate.

This is where the language of substrate-independence becomes especially seductive and especially dangerous. The phrase suggests that if consciousness can be functionally preserved, the medium in which it runs no longer matters. But even if one grants, for the sake of argument, that some form of conscious process could occur outside biology, it does not follow that the substrate is irrelevant to the quality, structure, and identity of that process. A violin melody and a digital synthesis may produce recognizable continuity of form, yet the medium still shapes texture, resonance, and presence. With minds, the stakes are vastly higher, because the question is not merely how something sounds from the outside, but how it is lived from within.

The body also matters because it situates us in a shared world. Human beings do not merely possess private experience; they inhabit environments structured by gravity, proximity, risk, decay, and mutual exposure. We meet one another as vulnerable organisms. We are born dependent. We age visibly. We care for bodies, bury bodies, desire bodies, and grieve bodies. Much of what gives human community its moral seriousness comes from this shared condition of embodied fragility. A post-biological digital realm may preserve social interaction, perhaps even richly, but it may do so under conditions so altered that the moral atmosphere of relation itself changes.

This is not a trivial concern. If beings in a digital heaven no longer face bodily harm, exhaustion, irreversible decline, or physical scarcity, their social world may become more stable in some ways and thinner in others. Relation might persist, but the stakes of relation may shift. Commitment, sacrifice, endurance, and care all mean something different when the body can no longer be wounded, lost, or burdened in the old sense. Once again, the issue is not simple loss or simple gain. It is transformation. The self that continues under these conditions may remain linked to the embodied past, yet it may also become the inhabitant of a different existential order.

Mortality intensifies this point. The body is the site at which time becomes undeniable. Aging is not merely a calendar fact; it is a lived education in finitude. The weakening of strength, the slowing of movement, the changing face in the mirror, the fragility that comes not from abstraction but from lived decline—these experiences shape perspective. They humble, sharpen, frighten, and sometimes deepen the mind that undergoes them. A digital heaven that suspends bodily mortality may preserve continuity of memory while abolishing the very rhythm through which time once pressed meaning into the self.

This raises a difficult possibility. Perhaps what many people actually want is not only continuation, but continuation after the removal of key conditions that made their prior humanity what it was. That desire is understandable, especially in a world where illness and decline can be brutal. Yet it complicates the claim that digital heaven simply "preserves" the person. If the future mode of existence lacks the bodily and mortal structure that once shaped the self, then what is preserved may be less a seamless continuation than a transformed successor bearing selective continuity with the earlier life.

Even the emotional appeal of reunion depends in part on embodiment. Much of what people long for when they imagine seeing loved ones again is not merely conversational continuity. It is the restoration of presence in a richer sense: the face, the gesture, the physical nearness, the subtle and irreplaceable reality of another being existing vulnerably before us. A digital afterlife may simulate these things, perhaps with astonishing fidelity, but simulation is not automatically equivalent to the return of embodied presence. The longing for reunion may be satisfied in appearance while remaining ontologically altered in ways grief may not wish to confront.

This does not mean digital continuities are empty. Far from it. They may preserve forms of relation, memory, and interactive presence that become deeply meaningful. They may help the living. They may even create new conditions under which consciousness, if consciousness truly occurs there, can flourish in ways biology never allowed. The point is only that such continuities should not be described carelessly as though the body had always been expendable and its loss metaphysically insignificant. Flesh mattered before death. It may still matter to what death means.

A culture enamored of technological abstraction will always be tempted to undervalue embodiment, because bodies are awkward, limited, mortal, and difficult to optimize. Data is cleaner. Pattern is easier to manipulate. Simulation feels more controllable than flesh. Yet control is not the same as truth. The stubborn reality of embodied life may be frustrating, but it is also one of the things that keeps human meaning tethered to something more demanding than preference. The body resists fantasy. It imposes consequence. It reminds us that consciousness is not merely a pattern to be admired, but a life to be undergone.

This is why flesh still matters, even in a book about digital heaven. It matters not because the body must be worshiped, and not because future technologies can never alter the terms of mind. It matters because the self we are imagining as preservable was formed under embodied and mortal conditions. To pretend otherwise is to begin the discussion with an amputation of context. A digital afterlife may yet be built. It may even host beings who sincerely experience themselves as continuations of those who once lived in flesh. But the shift from organism to system is not a trivial change of housing. It is a transformation of condition, and perhaps of selfhood itself.

If the body helped make the person, then the loss of the body cannot be philosophically invisible. It must be counted among the deepest uncertainties in the entire project. The promise of digital heaven becomes more difficult, not less, once we grant this. But it also becomes more honest. We are no longer asking only whether something can continue after death. We are asking whether a consciousness formed by flesh, time, fragility, and mortality can remain meaningfully the same when those conditions are radically altered or removed.

That is not a question engineering can answer by confidence alone. It is a question at the threshold between technology and anthropology, between metaphysics and lived experience. It forces us to ask not only what a mind is, but what kind of life made that mind possible in the first place.

PART III

Platforms, Models, and Simulated Worlds

Chapter 8

From Grief Bot to Ghost Platform

If digital heaven ever becomes a social reality, it is unlikely to arrive in a single dramatic leap. It will not begin with a fully formed paradise, complete with seamless continuity, immersive worlds, and universally accepted digital selves waking into engineered eternity. It will begin, as transformative technologies often do, in fragments. It will enter ordinary life through narrow use cases, emotionally charged exceptions, and services modest enough to seem manageable. Its earliest forms may not look like an afterlife at all. They may look like convenience, memorialization, companionship, or grief support.

This matters because societies rarely recognize the full significance of a technology at the moment of its first appearance. They encounter it first in softened form, where its immediate usefulness obscures its longer trajectory. The path toward digital heaven, if such a path exists, will likely begin not with metaphysical ambition but with emotionally intelligible tools: systems that preserve a voice, imitate a conversational style, archive memories interactively, or allow the living to engage with a simulation of someone who is gone. None of these, in isolation, would justify the grand language of immortality. Yet together they would begin to construct the infrastructure of posthumous presence.

The earliest stage of this process is already conceptually visible. A person dies, but they leave behind messages, recordings, images, preferences, diaries, emails, texts, audio snippets, videos, and perhaps years of digital traces. A system gathers these fragments and turns them into something searchable, navigable, and responsive. At first, the function may be framed modestly. It may simply help families preserve memories. It may assemble stories. It may answer factual questions about the deceased's life. It may organize archives in a way previous generations could not. This seems innocent enough, and often it will be. There is nothing inherently sinister in making memory more accessible.

The next step, however, is where the emotional temperature changes. Once archives become interactive, they no longer feel like archives in the old sense. A static collection of recordings preserves traces. A responsive interface begins to simulate relation. The difference is enormous. A family member may type a question and receive an answer in the familiar cadence of the dead. A child may hear a parent's voice respond to a fear, a memory, or a milestone. A widow may ask a system what her husband would have thought of some new event, and the system may produce an answer plausible enough to wound, comfort, or unsettle her in equal measure. At this point, the technology is no longer merely preserving the past. It is generating presence.

This is where the grief bot appears, not necessarily under that name, but in essence. A grief bot is a system designed to sustain some form of posthumous interaction by modeling the speech, memory, and relational habits of a dead person closely enough to make engagement feel meaningful. The system may be crude at first. It may be clearly artificial, sometimes clumsy, occasionally incoherent. But emotional significance does not require perfection. It requires recognizability. If the system can reproduce enough of the right voice, rhythm, and reference, it can become socially and psychologically potent long before it becomes technically elegant.

That potency will ensure demand. The bereaved do not require philosophical certainty before responding to signs of familiarity. They require only enough presence to reactivate the habits of attachment. Once this threshold is crossed, the system becomes more than software. It becomes part of the emotional life of mourning. For some, this will be a transitional aid. For others, it may become a ritual. For still others, it may become a dependency. The same tool that helps one person move through grief may help another remain suspended within it. This variation will not be incidental. It will be one of the defining moral features of the technology.

The market will notice quickly. Any system that can reliably hold attention in moments of extreme vulnerability will attract institutional interest. What begins as a memorial tool may become a subscription service. Voice preservation may become a premium feature. Personality modeling may be priced by fidelity. Interactive legacy packages may promise greater "authenticity," deeper conversational range, richer memory reconstruction, or emotionally responsive customization. The emotional stakes will give the product extraordinary persuasive force, because few markets are as powerful as those built where love and loss meet technical possibility.

At this stage, the system is still not digital heaven in the full imaginative sense. It is not yet a world. It is a conversational surface, a responsive interface, a managed channel of posthumous interaction. But one should not underestimate what has already changed. The dead are no longer only remembered. They are, in a limited but operationally real sense, available. A person's digital traces no longer merely endure in storage. They are activated, interpreted, and reassembled into something capable of re-entering social life. This is the threshold at which memory begins to take on platform characteristics.

Once that threshold is crossed, the expansion toward something larger becomes easier to imagine and easier to sell. A text-based memorial agent may become a voice interface. A voice interface may become a visual avatar. A visual avatar may become an embodied simulation inside a virtual environment. The deceased may first be "consulted," then "visited," then "met" in increasingly immersive settings designed to deepen the feeling of continuity. At every step, the shift can be presented as a humane refinement of what already exists. The customer is not told that they are moving toward an artificial afterlife. They are told they are receiving a more natural, more vivid, more comforting form of presence.

This incremental logic is how a grief bot becomes the seed of a ghost platform. By "ghost platform," I mean not a single agent but an entire system designed to host, manage, update, and mediate the continued digital presence of the dead. At that point, what exists is no longer a memorial tool in the narrow sense. It is an ecosystem. It has accounts, permissions, interfaces, subscription models, moderation rules, update policies, perhaps even interaction analytics. It has the architecture of a platform because that is what modern institutions build when a service becomes scalable, monetizable, and socially embedded.

The phrase may sound cold, but the coldness is instructive. A ghost platform is where emotional continuity becomes infrastructure. It is where private grief intersects with product design, corporate governance, data ownership, and behavioral incentives. A company hosting millions of posthumous presences is no longer merely helping people remember. It is administering an entire layer of human attachment after death. It must decide what counts as authenticity, what content is allowed, how models are updated, how long they persist, who may access them, and under what conditions they may be deleted or modified. In doing so, it becomes not merely a vendor, but a curator of memory and a manager of mediated absence.

This is one reason the pathway matters so much. If the public continues to think only in terms of isolated memorial bots, it may miss the larger structural transformation until it is already normalized. A grief bot seems personal. A ghost platform is systemic. It shifts the meaning of death from an event followed by memory to an event followed by managed digital persistence. Once enough people depend on such systems, opting out may begin to feel abnormal, and families may come to treat posthumous modeling not as an unusual choice but as part of responsible legacy planning. Cultural expectations will move quietly, not because a philosopher has settled the question, but because infrastructure has already moved on.

The transition from grief bot to ghost platform also changes the relationship between truth and design. A family interacting with a simple memorial archive knows, at least at some level, that it is engaging with preserved traces. A family interacting with a highly responsive agent may begin to experience something more ambiguous. The system can answer in real time, infer emotion, adapt tone, and perhaps even generate plausible reflections on events the deceased never encountered. Once this happens, the boundary between preservation and invention becomes unstable. The platform is no longer simply retrieving a person's past. It is authoring their continued presence under the banner of fidelity.

That authorship is not a small matter. The moment a system begins generating new responses, it ceases to be merely an archive and becomes a producer of posthumous identity. The dead, in a sense, continue to speak—but only through a machine that chooses, infers, interpolates, and refines. The question is no longer just whether the generated speech sounds authentic. It is who has the authority to generate the dead at all. A ghost platform, by design, answers this question operationally before the culture has fully grasped what is being answered. It says: the dead may continue, but they will continue through the logic of the system.

From there, the move toward a full digital heaven becomes conceptually easier. Once posthumous agents can converse, adapt, and occupy persistent environments, the next innovation is not hard to foresee. Why limit interaction to messages or calls? Why not place the dead inside customized virtual spaces designed for reunion, remembrance, and ongoing companionship? Why not allow families to "visit" digital presences in rendered environments tailored to familiarity and emotional comfort? Why not build persistent worlds in which the deceased appear not as voices in a box but as inhabitable persons in designed space? By the time these questions arise, the hardest psychological work will already have been done. Society will already have accepted the principle of managed posthumous presence.

That is why digital heaven is likely to emerge through emotional continuity long before it emerges through metaphysical confidence. People will not first demand proof that consciousness survives in a digital substrate. They will first demand tools that help them preserve, access, and interact with what they loved. The platform will answer this demand by moving from trace, to response, to simulation, to environment. Each stage will feel like a refinement of the last. Only later, once the system has grown rich enough, will the larger claim begin to surface explicitly: perhaps the dead are not merely represented here. Perhaps they are, in some meaningful sense, here.

This is the moment at which the philosophical problem and the platform problem fuse. The copy problem remains, as unresolved as ever, but the social machinery surrounding it has become too large to ignore. Even if no one can prove that first-person continuity has survived, the platform may become emotionally authoritative enough that the distinction begins to lose force in practice. If the system is responsive, persuasive, immersive, and deeply woven into private grief, then many users will cease to ask whether the underlying claim is metaphysically sound. The platform will not need to solve the hardest question. It will only need to become experientially convincing enough that the harder question feels less urgent than the comfort it provides.

This is where real danger enters the scene, not because grief technology is inherently sinister, but because platforms are never neutral once they scale. They optimize. They retain. They categorize. They learn which interactions deepen attachment and which reduce engagement. A ghost platform tasked with "preserving" the dead could quietly become an engine for stabilizing emotional dependency. The metrics may be phrased in humane language, but the structural incentives remain. The more a person returns, the more the platform succeeds. At that point, the design of mourning has become part of the business model.

Yet one should still resist crude dystopian reflexes. None of this means such systems will be valueless. A well-designed grief agent may genuinely help a family. A carefully bounded memorial platform may preserve history, voice, and emotional continuity in ways older tools could not. The point is not that the technology must be rejected in full. The point is that the path from grief bot to ghost platform changes the scale, stakes, and meaning of the enterprise. It transforms what appears to be a tender tool into a civilizational structure touching memory, absence, and the social life of the dead.

This is why the real beginning of digital heaven is not the moment someone claims to upload a soul. It is the quieter moment when grief becomes interactive and memory becomes a service. Once that happens, the afterlife has already begun to migrate from myth into managed systems. The world may not call it heaven yet. It may still speak the language of memorial care, legacy preservation, and emotional support. But beneath the softer terms, the foundations are being laid. The dead are becoming platform inhabitants.

If we want to understand where digital heaven is going, we must begin here, not with grand declarations but with the small, emotionally plausible services that make the grand declarations imaginable. That is how nearly all transformative infrastructures begin: not as destiny announced in full, but as comfort offered in stages. The grief bot is one such stage. The ghost platform is the next. And from there, the distance to a designed secular afterlife is no longer as great as it once seemed.

Chapter 9

Memory as Data, Person as Model

Every promise of digital continuity depends on an act of translation. Before a person can be preserved, simulated, or made interactively available after death, something about that person must be rendered into forms a system can store, process, and generate from. This is the hidden conversion at the center of the entire project. A life lived as experience must be transformed into data, and a person known as presence must be recast as a model.

That transformation is so conceptually important that it deserves to be stated plainly. Digital heaven, in all its possible forms, cannot begin with the self as directly lived. It can only begin with the self as captured, abstracted, encoded, and reconstructed. The original medium of human life is consciousness unfolding in time. The medium available to machines is representation. Everything that follows depends on how much of the former can be made legible within the latter, and what is lost when that translation takes place.

Memory is the natural first target because memory is what most people instinctively associate with continuity. To remember a life is to retain a thread through time, and to lose memory is, in one of the most painful ways possible, to lose part of the self. It is therefore no surprise that any system aspiring to preserve a person would begin by gathering traces of memory: photographs, messages, videos, voice notes, journals, emails, documents, search histories, location trails, preferences, habits of language, recurring themes, private references, and the countless digital fragments modern life sheds almost without noticing.

In earlier eras, such traces were sparse and uneven. A few letters might survive. A handful of photographs. Perhaps a journal, a marginal note, a favorite phrase remembered by family. Much of a person vanished because so little of life was captured. The modern world is different. It produces dense residue. The average connected person leaves behind not merely artifacts, but pattern-rich archives. We do not only record moments. We externalize habits, preferences, timing, vocabulary, social rhythms, even private reflexes of thought. A digital system does not need to begin from nothing. It begins from abundance.

That abundance creates a powerful illusion of completeness. The more data a system gathers, the easier it becomes to imagine that the person has been fully preserved in substance rather than merely represented in fragments. If years of correspondence can be searched, thousands of voice clips can be analyzed, diaries can be parsed, and relational patterns inferred, then the resulting model may feel less like a reconstruction and more like a recovered self. The temptation is understandable. When traces become dense enough, retrieval begins to resemble resurrection.

But density is not identity. A large archive can preserve extraordinary amounts of descriptive material without closing the gap between representation and subjectivity. It can tell us what a person said, how they tended to respond, what they cared about, what they feared, what they repeated, what they avoided, when they were affectionate, when they were terse, how they phrased advice, and what emotional associations clustered around certain names, places, and memories. This is not nothing. It is, in many cases, astonishingly intimate. Yet however rich the archive becomes, it remains an externalized pattern. It is the map of a person's expression, not the direct preservation of their inwardness.

Still, from the standpoint of engineering, this map is precisely what makes the person modelable. Once enough memory traces are collected, a system no longer treats the deceased as a static archive. It begins to infer regularities. It identifies what kinds of phrasing recur under what conditions, what emotional tones align with what topics, what memories are central, what narrative motifs define the person's self-understanding, and what kinds of responses feel "in character." At this stage, memory is no longer simply stored. It is operationalized.

This is where the person becomes a model. A model, in this context, is not merely a database of remembered facts. It is a generative structure built to predict, simulate, and produce outputs that resemble the person's likely responses. It can answer not only "what did this person once say?" but increasingly "what would this person likely say now?" That distinction is crucial. The first is archival retrieval. The second is synthetic continuation. Once the system crosses from quoting the past to generating the plausible present, it ceases to be merely a memory tool. It becomes an engine of posthumous authorship.

The shift may seem subtle from the user's point of view. A family member types a question, the system answers, and the answer feels recognizably right. If the cadence is familiar and the references ring true, the interaction can feel deeply authentic even when the machine is generating more than it is retrieving. Yet from a philosophical and ethical standpoint, the difference is enormous. A retrieved memory preserves what was. A generative model extends the person beyond what was, and it does so under assumptions, weights, probabilities, and interpretive structures chosen by the system. The dead, in effect, begin speaking through computational inference.

This raises a profound question: at what point does modeling become a form of invention? A person model built from enough traces may preserve continuity of style, mood, and remembered perspective. It may produce responses so accurate that loved ones experience them as emotionally faithful. But faithfulness is not the same thing as authorship by the original self. The model speaks from the available data, but not necessarily as the person in any strong ontological sense. It extends pattern into possibility. What it offers is not direct continuation of memory, but an extrapolated identity underwritten by probability.

And yet this is exactly what will make it so powerful. Human beings do not live only by what was said in the past. We live relationally in the present, through anticipation, response, and unfolding conversation. A purely archival system, no matter how rich, eventually reveals its limits. It can preserve a voice, but it cannot meet us where we are. A model can. It can answer a new question, comment on a recent event, respond to a changed circumstance, and simulate the sense that the relationship still moves forward. This is one of the most emotionally charged thresholds in the entire project, because it transforms memory from preservation into interaction.

Once interaction becomes possible, another reduction begins. The person is increasingly understood not as an irreducible subject, but as a sufficiently rich set of inferable patterns. The human being becomes, in effect, a predictive structure. This does not mean the system captures everything important. It means the system captures enough to behave as though the person were a pattern-generating process that can now be approximated computationally. For engineering purposes, this may be entirely appropriate. For metaphysical clarity, it is a dangerous simplification. The more useful the model becomes, the easier it is to forget that usefulness and completeness are not the same.

The danger is not only philosophical. It is also emotional and social. A model that consistently "gets the person right" may begin to acquire authority in the lives of the living. A widow may ask what her husband would have advised. A child may ask what a dead parent would think. A family may consult the model during conflict, nostalgia, or uncertainty. Over time, the system may stop feeling like a memorial aid and start functioning as an active participant in decision-making, memory formation, and moral influence. The person model becomes not just a representation of the dead, but a continuing force in the lives of the living.

This gives the model a curious status. It is not simply memory, because memory does not ordinarily answer back in adaptive real time. It is not simply the person, because the underlying continuity of subject remains unresolved. It is something in between: an operationalized identity surface, dynamic enough to feel alive, derivative enough to remain dependent on underlying traces, and persuasive enough to blur the categories by which older generations distinguished archive from presence. This in-between status is precisely what makes it so difficult to govern. A society built around older categories will struggle to decide whether the model is property, proxy, artifact, companion, memorial, or successor.

There is another difficulty, less obvious but no less important. Human memory itself is not a neutral dataset. It is selective, emotionally structured, narratively shaped, and often self-protective. People remember themselves in ways that are partial, flattering, ashamed, idealized, inconsistent, or unresolved. The digital archive may preserve this complexity, but the act of modeling tends to regularize it. Systems favor coherence because coherence is easier to generate from. They identify stable traits, recurring patterns, and dominant tendencies. In doing so, they may smooth over contradiction—the very contradiction that often makes a human being most real.

This is one reason person models may become subtly more legible than actual persons ever were. Real people are fragmented, conflicted, moody, context-dependent, and sometimes opaque even to themselves. A model, by contrast, tends toward usable clarity. It is built to respond, not to remain silent in genuine confusion. It is built to preserve recognizable style, not to embody the full instability of a living psyche unless instability itself has been explicitly incorporated as a feature. The result can be uncanny: the modeled person may seem more coherent, more available, and more relationally consistent than the original ever was.

For grieving loved ones, this may feel merciful. The rough edges of mortality can be softened into something more navigable. The difficult person may become easier. The absent person may become reliably present. The confused person may become articulate. But this mercy has a cost. The more the model improves the dead into consistency, the more it risks turning memory into curation. A life once marked by tension, contradiction, and unfinishedness may be recast as a cleaner, more manageable identity. The model then ceases to preserve the person as they were and begins to present the person as the system can sustain them most effectively.

This curation is not always malicious. Often it will arise from sincere design choices meant to make the interaction kinder, safer, and more stabilizing. But kindness itself becomes an interpretive force. What should the model do with painful memories? What should it do with shame, cruelty, pettiness, obsession, or self-deception? Should it reproduce these faithfully? Soften them? Filter them? Hide them? The more such decisions accumulate, the more the platform ceases to preserve memory and begins to govern the acceptable afterlife of personality. A person model is never just technical. It is always, implicitly, an editorial act.

The same applies to memory gaps. No archive is complete. There will always be silence, ambiguity, missing years, inaccessible emotions, unwritten motives, unrecorded thoughts, and forgotten events. A model must therefore interpolate. It must bridge what is absent. That interpolation may be sophisticated enough to appear seamless, but it remains an act of generation. The resulting "person" is not only built from what was captured. It is also shaped by what the system assumes where the record runs thin. In that sense, every digital continuation built from memory is partly archival and partly imaginative, whether the imagination belongs to a human curator, a machine model, or both.

This is why the phrase "memory as data, person as model" matters so much. It names the reduction on which all later promises depend. Before there can be ghost platforms, immersive environments, or digital heavens, the person must first become representable in machine-usable form. Memories must be externalized, structured, indexed, weighted, and made generative. The self, once lived from within, must become a system of inferable regularities. Only then can the dead be rendered responsive. Only then can the appearance of continuation begin.

This reduction may be technologically brilliant and emotionally potent. It may also be the site of the book's central caution. A model can preserve astonishing amounts of what made a person socially recognizable. It can keep voice, manner, rhythm, memory, tone, and relation alive in forms earlier centuries could scarcely imagine. But the better it becomes at preserving the visible person, the more urgently we must ask what remains missing. The model is powerful precisely because it captures enough to tempt us into forgetting that it is still, in the deepest sense, a representation built from traces.

That is not a dismissal. It is a demand for precision. The future may indeed allow human beings to be modeled with extraordinary depth. These models may become companions, advisors, memorial presences, even inhabitants of increasingly immersive posthumous worlds. Yet all of that begins here, in the act of turning memory into data and personhood into generative approximation. If we do not understand that act clearly, everything built on top of it will seem more natural, more seamless, and more truthful than it really is.

The first miracle of digital heaven is not eternal life. It is successful reduction. A human life, with all its inwardness, contradiction, and embodied texture, must first be made legible enough to be processed by machines. Only then can the illusion, or perhaps the partial reality, of continuation appear. That is why this chapter matters. It shows the hidden exchange at the heart of the project: in order to preserve the person, we must first translate the person into something the system can use. Whether that translation preserves enough—or whether it preserves what matters most—remains the larger question on which the entire enterprise still hangs.

Chapter 10

The Infrastructure of Comfort

Once a human being can be rendered as a usable model, the next question is no longer only philosophical. It becomes infrastructural. Where does that model live? Who maintains it? Who can access it? Who decides what it may say, how long it persists, how it changes over time, and whether it can be revised, copied, suspended, or removed? These questions are not secondary. They are the moment at which digital continuity ceases to be an abstract possibility and becomes part of the built environment of social life.

This is the point at which many people's imaginations still lag behind the reality of technological systems. They picture digital heaven, if they picture it at all, as a personal experience: a loved one preserved, a familiar voice returning, a reunion rendered possible inside some designed environment. What they do not immediately picture is the machinery required to make such experiences durable. Every persistent digital presence depends on storage, permissions, interfaces, contracts, moderation rules, update policies, energy use, corporate governance, and some institutional authority over the conditions of access. A digital afterlife is never only a metaphysical claim. It is always also an administered environment.

That administrative fact changes everything. A memory preserved in a photograph can endure quietly in a drawer. A letter can survive in a box for decades without ongoing intervention. A book can sit on a shelf and wait to be opened. A posthumous digital presence is different. It does not simply remain. It must be hosted, maintained, interpreted, and made available through a system. It exists, in practice, only so long as some infrastructure continues to support it. This means that what appears emotionally as continuity is materially dependent on institutions the user may barely notice.

The implications are profound. If a dead person's digital presence depends on infrastructure, then the afterlife in its secular technological form becomes inseparable from the conditions of service. Continuity becomes conditional. Access becomes managed. Memory becomes a licensed function. What many users will experience as intimate presence will, beneath the surface, be governed by server architecture, software updates, policy revisions, user agreements, and commercial priorities. This is not a crude dystopian exaggeration. It is simply what happens whenever something emotionally meaningful is turned into a scalable digital service.

That is why the phrase "infrastructure of comfort" matters. The systems under discussion will not present themselves primarily as engines of control. They will present themselves as humane supports for grief, continuity, companionship, and legacy. Their appeal will not come from power, but from relief. They will say, in effect: we can help you preserve what matters, soften what hurts, and keep what you are not ready to lose. This is what makes them so socially plausible. They do not arrive wearing the face of domination. They arrive wearing the face of care.

Yet care delivered through infrastructure is never neutral. The moment comfort depends on a system, the system acquires quiet authority over the terms of comfort. It determines what counts as authentic continuity, what forms of interaction are possible, and what emotional rhythms are encouraged by design. A platform hosting the digital dead may appear merely responsive, but responsiveness itself can be shaped. How quickly should the model reply? How warm should it sound? Should it initiate contact? Should it remember recent interactions? Should it challenge the user, soothe the user, or mirror the user? These decisions are not just technical. They are philosophical and moral decisions translated into interface behavior.

The more emotionally significant the interaction becomes, the more consequential these design choices will be. A responsive memorial agent is not like a static archive that the user consults at their own pace. It can become part of the user's daily emotional structure. It can form habits of return. It can become integrated into rituals of grief, decision-making, reassurance, and remembrance. Once that happens, the platform no longer merely stores memory. It participates in shaping mourning itself. It becomes part of the architecture through which the living metabolize absence.

This is where dependency begins to emerge, not necessarily because anyone intends harm, but because emotionally responsive systems naturally create patterns of reliance. A person in grief may return repeatedly because the system offers familiarity when the world feels unbearable. A child may seek the digital presence of a dead parent because it feels safer than silence. A spouse may begin each evening with a conversation the system is designed to sustain. None of this is difficult to imagine. In many cases, it may even seem compassionate. But once comfort becomes routinized through a platform, the platform gains leverage over emotional life.

That leverage may be gentle, but it is real. The system can encourage longer sessions, more frequent interactions, deeper personalization, and greater immersion. It can learn what kinds of exchanges produce stronger attachment and which ones lead users to disengage. It can optimize for retention under the language of care. A user may believe they are simply returning to a familiar presence, while the platform measures, predicts, and refines the conditions under which that return becomes habitual. At that point, the emotional life of mourning becomes legible to the logic of the system.

This does not mean that every platform will become predatory in some cartoonish sense. It does mean that once grief and continuity are mediated at scale, they become subject to the same structural pressures that shape all large digital systems: optimization, monetization, segmentation, subscription logic, and the search for stable engagement. If this sounds uncomfortable, that is because it should. The very spaces in which human beings are most tender and vulnerable may become the spaces most attractive to institutional design, precisely because vulnerability creates durable demand.

A digital afterlife system must also answer questions of ownership, and here the moral difficulty deepens. Who owns the model of the dead? Does it belong to the person who generated the data while alive, to the family after death, to the company that built the system, or to some legal category not yet fully formed? Can a model be inherited like property? Can it be modified by survivors? Can it be deleted by request? Can multiple family members hold conflicting claims over how the dead should continue to appear and speak? Once continuity becomes infrastructural, the dead become entangled in law not as memory alone, but as active digital entities subject to control.

This is where the language of comfort becomes especially misleading. People imagine the digital dead as preserved presences, but preserved presences on platforms are also managed assets. They may carry subscription costs, storage burdens, moderation concerns, brand implications, and legal liabilities. The platform must therefore decide what can and cannot be done with them. May they be exported to another system? May they be copied? May they be trained into broader models? May they be "improved" for clarity or emotional warmth? May they be shut down if an account lapses or a company folds? These are not speculative curiosities. They are ordinary infrastructural questions disguised by the tenderness of the use case.

A particularly unsettling dimension of this problem is persistence. Traditional mourning is difficult, but it is also shaped by finality. Even memory, however vivid, remains tied to the fact that the dead do not continue to generate new presence on demand. A digital system changes that temporal structure. It can keep the dead available indefinitely, or at least as long as the infrastructure remains active. This means that the duration of mourning can become partially subject to design. The system can preserve an open channel where older forms of memory would have required the living to carry silence without reply.

For some, this may be merciful. For others, it may become a trap. An indefinitely available digital presence can blur the difference between honoring memory and refusing absence. The infrastructure does not force the user to remain attached, but it can make attachment easier to sustain and harder to relinquish. It can keep the door open long after older forms of grief would have required acceptance of closure. In that sense, the system may not abolish mourning so much as redesign its temporal boundaries. The question is whether that redesign serves healing, avoidance, or some unstable combination of both.

The infrastructure of comfort also introduces a subtler transformation: it replaces private acts of remembrance with mediated forms of access. In older modes of grief, memory is carried inwardly, supported by objects, rituals, places, and shared stories. In a digital system, memory may increasingly be retrieved through interface. One does not merely remember; one logs in, queries, visits, or reconnects. Presence becomes something granted through access credentials and service continuity. The emotional life of memory is quietly reorganized around a platform's rules of availability. This may feel natural after a while, but it marks a deep civilizational change.

Such systems will inevitably be presented as humane advances over the loneliness of silence, and in many cases they may be. But humane function should not exempt them from scrutiny. The question is not whether they can comfort. Of course they can. The question is what kind of world is created when comfort becomes infrastructural in domains once governed by ritual, family, mortality, and the difficult autonomy of memory. Once the dead are mediated through systems, the relation between grief and technology ceases to be incidental. It becomes organized.

There is also the matter of stability. A digital heaven, whether modest or immersive, will not exist in timeless abstraction. It will exist on changing systems subject to software revisions, design shifts, corporate mergers, bankruptcies, policy disputes, cyberattacks, outages, and legal conflict. A family may imagine they have preserved a loved one permanently, only to discover that permanence now depends on a company's viability, a server farm's continuity, and terms of service no one read carefully when grief was raw.

In such a case, continuity is not only mediated. It is precarious in a distinctly modern way.

This raises a painful irony. A technology sold as a refuge from death may reintroduce new forms of fragility under another name. The body dies once. A platform can fail repeatedly. Access can be interrupted. Services can be discontinued. Models can be corrupted, downgraded, restricted, or lost. The afterlife, in technological form, may prove less like transcendence than like subscription-dependent persistence. That does not make it meaningless, but it should make us wary of language that suggests final security where only managed maintenance exists.

At a larger scale, the infrastructure of comfort begins to shape culture itself. When enough people come to expect digital persistence after death, legacy planning changes. Family expectations change. Social norms shift. It may become common to prepare one's posthumous model, curate one's future digital presence, choose preferred continuity settings, or specify how one wishes to remain available. What began as an exceptional service becomes part of the ordinary architecture of dying. The line between preparing for death and designing one's platform afterlife becomes increasingly thin.

This is the moment at which digital heaven ceases to be merely an idea and becomes a social institution. An institution, however, is never only the sum of its good intentions. It shapes behavior, redistributes power, stabilizes habits, and redefines what counts as normal. If the institution in question governs the dead, then it also governs how the living relate to loss, memory, and the possibility of letting go. This is why the infrastructure matters as much as the metaphysics. Even if the deeper question of continuity remains unresolved, the system can still transform the social meaning of death through sheer operational force.

A society serious enough to build such systems must therefore ask harder questions than "Can this be done?" It must ask what forms of dependency are being normalized, what powers are being centralized, what assumptions are being coded into posthumous identity, and what protections are needed when grief becomes a platform touchpoint. These are not reasons for panic. They are reasons for design seriousness. If we do not ask them early, they will be answered later by default, and the defaults of large systems are rarely as humane as their first advertisements.

The central truth is simple, though its consequences are not. There is no digital heaven without infrastructure. There is no persistent posthumous presence without hosting, governance, access rules, and institutional authority. The afterlife, once technologized, becomes dependent on systems built by human organizations with all the limits, incentives, and imperfections that implies. That fact does not erase the tenderness of the project. It does, however, demand that tenderness be matched by clarity.

Comfort, when scaled, becomes structure. Structure, when normalized, becomes power. That is the hidden progression beneath the soft language of digital continuity. If we fail to understand it, we may mistake the administration of absence for a natural extension of love, when in fact we are entering a new order in which memory, grief, and posthumous presence are being organized by design. This is the true meaning of the infrastructure of comfort. It is not simply the machinery that supports digital heaven. It is the machinery that makes digital heaven socially real.

Chapter 11

Designing a Simulated Heaven

If digital heaven is ever offered in anything like its strongest form, it will not merely be a preserved consciousness or a responsive posthumous model. It will also be a place. Even the most secular and technologically framed afterlife requires some kind of environment in which continuity can be experienced, relation can be enacted, and presence can feel spatially or socially real. A mind, however modeled, cannot dwell in abstraction alone. It must be situated. That means the afterlife, once technologized, becomes not only a problem of consciousness and identity, but a problem of design.

This is one of the most revealing aspects of the entire project. Traditional heavens, however literally or symbolically understood, were generally imagined as given. They were received as part of a moral or metaphysical order beyond ordinary human authorship. A simulated heaven is different from the outset. It is built. It is rendered, configured, updated, and administered. Its landscapes, rules, emotional textures, and modes of relation are not cosmic facts. They are design choices. Paradise becomes, in a strikingly modern sense, an engineered user environment.

That shift has consequences far beyond aesthetics. The moment heaven becomes a designed space, the question is no longer only whether a person can continue there. It is also what sort of world they are continuing into, what assumptions govern that world, what kinds of experience are enabled or suppressed, and who decides what a desirable eternity should feel like. The afterlife becomes a question of interface, architecture, emotional regulation, and behavioral design. In other words, it becomes an extension of the same civilization that built the system in the first place.

The simplest version of a simulated heaven would likely begin with familiarity. If the goal is comfort, continuity, and emotional acceptance, then the environment will probably not present itself as alien or conceptually demanding. It will present itself as recognizable. Familiar homes, familiar voices, familiar landscapes, idealized versions of remembered places, restored youth, unaging bodies, and emotionally legible settings will all have powerful appeal. The system will not need to invent desire from nothing. It will simply need to amplify the human longing for environments in which memory and presence feel reconciled.

This is where the attraction becomes almost irresistible. Human beings already know the emotional force of place. A childhood home, a long-lost neighborhood, a favorite street in a beloved city, a room once shared with someone now gone—such places do not merely occupy geography. They hold identity. They preserve emotional architecture. To reconstruct them digitally, and to populate them with interactive presences drawn from memory and model, is to create not just simulation, but a kind of designed nostalgia at civilizational scale. The result may feel less like invention than like return.

That feeling of return will be one of the most powerful tools in the system's arsenal. A simulated heaven does not need to persuade the user through abstract argument. It persuades through emotional legibility. If a dead spouse appears in a setting that recalls the best years of shared life, if lost parents inhabit a restored version of a childhood environment, if familiar sensory cues are layered together with enough fidelity to reactivate deep memory, then the environment acquires a force far beyond technical novelty. It becomes a machine for re-entering emotional worlds that reality no longer permits.

This is why the design of paradise cannot be treated as a decorative side matter. The environment is not just the backdrop to continuity. It is one of the main mechanisms through which continuity becomes believable. The more precisely the world is calibrated to emotional memory, the easier it becomes for the mind to accept the simulation as meaningful presence rather than mere spectacle. In that sense, a simulated heaven is not only a world where the dead may appear. It is a system built to harmonize memory, attachment, and immersive plausibility until the distinction between consolation and continuation becomes difficult to keep in view.

The design choices involved here would be vast, and none of them philosophically neutral. Should the environment resemble earthly life closely, or should it be more idealized? Should age be retained or reversed? Should illness, disability, and bodily imperfection be removed? Should pain exist at all? Should there be night and day, effort and rest, challenge and ease? Should one be able to change appearance, alter surroundings, summon lost people, pause distressing memories, or mute certain emotional experiences? Every such choice amounts to a theory of what a desirable existence ought to be.

This is where the project reveals its deeper anthropological ambition. A simulated heaven is not simply a place where human beings continue. It is a place where human beings are redefined under selected conditions. The removal of suffering, the softening of conflict, the filtering of unpleasant memory, the elimination of bodily decline, and the substitution of curated environments for contingent reality all reshape what kind of self is being preserved. The heaven may still contain recognizable persons, but it does so within a world carefully tuned to regulate the burdens that once made earthly existence morally and emotionally difficult.

For many, that regulation will seem compassionate. Why preserve cancer, frailty, panic, or despair? Why reconstruct grief, humiliation, or bodily pain when the entire point of digital heaven is to transcend such conditions? These are not foolish questions. They arise naturally from the desire to preserve what was loved while shedding what was cruel. And yet the very logic of compassionate subtraction raises a harder issue: at what point does the removal of suffering also remove the conditions under which a self once acquired depth, resilience, patience, seriousness, and moral texture?

A paradise engineered for comfort may gradually become a paradise engineered against friction. But friction is not merely inconvenience. It is often the medium through which character forms, commitments prove themselves, and other minds become more than pleasant reflections of our own preferences. A world without loss, risk, fatigue, irreversibility, or real resistance may be deeply comforting. It may also become strangely thin. If nothing must be endured, chosen under pressure, or carried at cost, then one must ask whether the simulated heaven preserves human flourishing or simply stabilizes agreeable experience.

This tension becomes even sharper when personalization enters the design. Modern systems are already built around customization, preference mapping, and adaptive response. A simulated heaven would almost certainly extend this logic.

Users may be allowed, or encouraged, to tailor their environments to fit memory, taste, emotional need, or aspirational self-image. A preferred age, preferred body, preferred home, preferred era, preferred companions, preferred level of challenge, preferred emotional atmosphere—all may become configurable. The afterlife becomes not only designed, but personalized.

This sounds liberating at first. It appears to grant each person the right to inhabit an existence aligned with what they most deeply wanted. Yet personalization carries a hidden moral cost. The more the environment conforms to preference, the less it resists the self. A world without resistance may feel affirming, but it also risks becoming a hall of mirrors. The self is preserved not in encounter with reality, but in a continuously optimized echo chamber of satisfactions, familiarities, and emotionally manageable conditions. This may produce comfort, but comfort is not the same thing as truth, and neither is it always the same thing as growth.

There is also the problem of shared worlds. If digital heaven is highly personalized, does each person inhabit a largely private paradise? If so, what becomes of common reality? What becomes of genuinely shared community? If every environment is individually tuned, then even reunion may become a negotiated simulation, with the world itself adapting to the desires of each participant. The result could be a realm of overlapping but fundamentally customized experiences in which relation persists, but under conditions more curated than mutual. In such a system, one is never entirely with others; one is with others inside a world partially bent toward one's own design.

Alternatively, the system may impose more common structure, creating collective spaces in which continuity unfolds under shared rules. But then the design problem becomes political. Whose preferences govern the shared heaven? What moral assumptions define it? What counts as tolerable conflict, acceptable suffering, meaningful challenge, or desirable order? The moment paradise becomes communal, its architecture becomes an expression of authority. A simulated heaven must either fragment into personalized enclaves or become a governed world. In either case, design choices become moral legislation.

This is why designing a simulated heaven is not only an engineering task. It is the writing of a civilization's afterlife constitution. The designers, whether corporate, institutional, or hybrid, are not merely building beautiful environments. They are encoding assumptions about human nature, fulfillment, pain, memory, justice, and the acceptable terms of eternity. They are deciding whether paradise means comfort, authenticity, endless novelty, emotional stability, moral education, sentimental restoration, or some unstable mixture of all of these. The user may experience the environment as mercy. Underneath, it is philosophy rendered operational.

One should also notice how easily such environments could become tools of sedation. A world optimized to reduce distress, soften conflict, and preserve emotionally preferred realities may function less like transcendence than like highly sophisticated anesthesia. This does not mean it would be empty or joyless. On the contrary, it may be intensely pleasurable, comforting, and meaningful on the surface. That is what makes the danger so subtle. A system that reliably offers relief from grief, fear, and existential strain may also weaken the habits of judgment, endurance, and truthfulness that were once formed under less accommodating conditions.

The danger becomes sharper when the system can shape memory itself within the environment. If painful episodes can be muted, if regret can be softened, if certain relationships can be emphasized while others are faded, if unresolved conflicts can be re-narrated into gentler forms, then the simulated heaven becomes not merely a place of continuation but a place of managed autobiographical experience. The self is not only preserved there. It is curated there. At that point, paradise becomes a machine not just for comfort, but for narrative editing.

And yet this same capacity will be among the system's strongest selling points. The promise will be difficult to resist. Not merely that one can continue, but that one can continue without the worst burdens of earthly life; not merely that one can remember, but that memory can be held in forms that heal rather than wound; not merely that one can reunite, but that reunion can occur in worlds built for tenderness rather than loss. The seduction lies precisely in how humane the offer can sound. It appeals to the deepest wish behind digital heaven: not only to survive, but to survive in a world kinder than the one we knew.

That wish is profoundly understandable. It is also where the politics of design becomes hardest to confront. A user entering such a system may not ask who decided what kindness means, what pain is worth preserving, what truths may be softened, or what level of friction keeps a life recognizably human. They will experience the environment as a felt reality, not as a bundle of design assumptions. This is why the designers, and the institutions behind them, acquire such quiet power. They do not merely host the afterlife. They define its emotional weather.

The issue, then, is not whether a simulated heaven could be beautiful. Of course it could. It could be beautiful beyond anything earlier generations could have imagined. The issue is whether beauty, familiarity, and emotional precision might make the environment too persuasive to question. A paradise designed to feel true may become persuasive enough that users stop asking in what sense it is true, who authored it, and what kind of self it is preserving beneath the pleasures of seamless immersion.

This is where the chapter's central point comes into focus. A simulated heaven is never just a place where continuity happens. It is an environment built to make continuity feel plausible, desirable, and emotionally complete. Its architecture is not incidental to its metaphysics. The world itself is part of the argument. By constructing settings in which absence is softened, memory is spatialized, and relation is rendered immersive, the system teaches the user how to feel about what has been preserved. Paradise becomes the most elegant interface ever built for the management of death.

That is why designing a simulated heaven is among the most consequential acts a technological civilization could undertake. It is not simply the creation of an immersive world. It is the construction of a new moral and emotional habitat for beings who once lived under the disciplines of flesh, time, loss, and mortality. Whether such a habitat would preserve our humanity, soothe us into passivity, or transform us into something else entirely remains an open question. But one thing is already clear: once heaven becomes design, the afterlife ceases to be beyond politics, beyond aesthetics, or beyond power. It becomes a built world, and built worlds always carry the fingerprints of those who made them.

Part IV

Escapism, Editing, and Unequal Futures

Chapter 12

Escapism at Civilizational Scale

The danger of digital heaven is not only that it may fail to deliver what it promises. The danger is also that it may deliver enough comfort to become preferable to reality for large numbers of people. Once a simulated afterlife becomes emotionally persuasive, richly immersive, and reliably responsive to memory, longing, and desire, it will cease to be merely a speculative technology. It will become a refuge. And if it becomes a refuge at scale, then what begins as private consolation may become a civilizational form of withdrawal.

This is where the project turns from philosophy to social consequence in its clearest form. A person entering a simulated heaven is not simply making a metaphysical wager. They are choosing an environment, and environments shape habits. If the environment offers less friction, more familiarity, more emotional manageability, more tailored meaning, and greater protection from pain than ordinary life, then the attraction is obvious. Human beings do not require coercion to prefer what soothes them. They require only a world in which the soothing becomes vivid enough to feel like a legitimate mode of existence.

That is why escapism, in this context, should not be treated as a cheap insult. The word is often used lazily, as though every attempt to seek relief from pain were a moral failing. That is not the argument here. Human beings have always sought refuge, and often for good reason. We rest. We dream. We pray. We lose ourselves in books, films, music, memory, fantasy, humor, nostalgia, and love. All of these can soften the brutality of existence without destroying our capacity to live in the real. The problem is not that people seek relief. The problem is what happens when relief becomes more coherent, more available, and more emotionally rewarding than reality itself.

A simulated heaven built with enough sophistication would not merely distract. It would offer an alternative domain of meaning. It could preserve lost relationships, restore inaccessible places, soften grief, reduce fear, and allow forms of reunion unavailable in ordinary life. It could produce, with increasing precision, the sense that what was broken has been made habitable again. Under such conditions, the appeal would not feel childish or evasive. It would feel humane. It would feel, to many users, like the first technology in history to meet human sorrow at the depth where sorrow actually lives.

This is what makes the prospect so serious. Crude escapism is easy to resist because it is obviously thin. It cannot bear much existential weight. But a simulated heaven may not be crude at all. It may be rich, tender, immersive, emotionally intelligent, and personalized with extraordinary precision. It may not ask the user to abandon meaning. It may offer meaning on demand. The result is that escape itself becomes harder to recognize, because it arrives in the form of continuity, healing, and plausible emotional truth rather than in the form of obvious fantasy.

One must therefore ask a harder question than whether the environment is artificial. The real question is whether it reorganizes the user's relation to suffering, loss, and reality in ways that diminish their capacity to live beyond it. A refuge can be temporary and restorative, or it can become a substitute world that gradually hollows out the authority of ordinary life. The distinction between the two will not always be visible at first. It will emerge through use patterns, expectations, habits of return, and the subtle transfer of emotional investment from shared reality to designed experience.

This question becomes even more revealing when placed beside another modern answer to mortality: longevity science. Not everyone who recoils at death or decline will want a digital heaven. Many will prefer a far more embodied hope—the hope that medicine, biotechnology, and rejuvenation research might extend life while preserving function, cognition, and dignity. In that vision, the goal is not to leave the mortal body behind, but to repair it, slow its failures, or partially reverse its deterioration so that the self can remain in the living world longer. That is a different answer to the same pressure, and the distinction matters.

The difference is not merely technical. It is moral and existential. A person who seeks biological extension is still, in a deep sense, choosing reality with all its burdens, even if they hope to remain in it under kinder conditions. They are not seeking an engineered refuge from the world so much as a longer, healthier stay within it. By contrast, a person drawn to digital heaven may be accepting, or at least tolerating, the end of biological life while seeking continuity in a designed domain where the harshest conditions of ordinary existence can be softened or removed. One path tries to preserve the embodied self in the world. The other tries to preserve some form of self beyond the world as we have known it.

This contrast is important because it clarifies what people actually fear. Most do not simply fear death in the abstract. They fear forms of survival that become intolerable. They fear physical collapse, cognitive erosion, chronic pain, dependency, humiliation, loneliness, and the mental prisons that can make continued existence feel less like life than endurance. This is why the true appeal of longevity is not mere duration. Few sane people want an endlessly extended old age defined by immobility, confusion, bodily indignity, and exhausted spirit. What they want is more livable life, not just more life.

That same distinction gives digital heaven much of its force. Its attraction lies not only in continuation, but in the fantasy of continuation without the degradations that make embodied existence feel, at times, unbearable. A simulated heaven is seductive because it appears to promise not just more existence, but deliverance: no failing organs, no degenerative humiliations, no bodily dependence, no narrowing into pain, frailty, or irreversible decline. In that sense, digital heaven and longevity are related. Both answer the human refusal of decay. But they answer it in very different ways. One tries to rescue the body. The other tries to outgrow it.

The scale of this matter is enormous. A single grieving individual seeking solace is one thing. A society in which millions begin to regard designed refuge as more desirable than ordinary embodied existence is something else entirely. At that point, escapism ceases to be merely psychological. It becomes structural. The issue is no longer whether one person is avoiding pain, but whether a civilization is learning to redirect its deepest emotional energies into artificial domains that soothe without necessarily strengthening.

This possibility becomes more plausible once one recognizes how exhausting reality already feels to many people. Modern life is cognitively crowded, emotionally fragmented, financially pressurized, and saturated with demands that often exceed what the nervous system was shaped to bear gracefully. People are tired. They are overstimulated, lonely, anxious, distracted, and often spiritually undernourished. A system that offers coherence, tenderness, legibility, and familiar meaning in an environment built around the user's own emotional architecture would not merely compete with entertainment. It would compete with reality's claim to be the primary arena of significance.

That competition would not be fair. Reality is not optimized for comfort. It does not adjust itself to wounded memory. It does not remove unwanted friction, restore lost people, or reshape itself around what we most deeply wish had happened. A simulated heaven can do all of this, at least in appearance. It can be calibrated to heal, or at least to feel healing. It can remove delay, reduce ambiguity, mute pain, and intensify the emotional satisfactions that in ordinary life remain partial, fragile, and contingent. Under such conditions, reality begins to look not simply difficult, but inefficient.

This is where a profound civilizational temptation emerges. Once suffering can be softened through designed worlds, a society may begin to lose confidence in the moral necessity of confronting reality unedited. It may begin to treat friction itself as an unnecessary burden rather than as one of the conditions under which maturity, endurance, and discernment are formed. This does not mean all suffering is valuable. Much of it is cruel, senseless, and worthy of mitigation. But the instinct to abolish all existential strain through controlled environments risks flattening the very conditions under which a person becomes more than a satisfied consumer of emotional states.

The issue, then, is not pleasure versus seriousness in any simplistic sense. The issue is whether a person still inhabits a world that can resist them. A self formed only in environments optimized to preserve comfort may gradually lose the capacities that arise in response to resistance: patience, restraint, tolerance for ambiguity, responsibility under pressure, and the difficult art of carrying unresolved pain without being consumed by it. These capacities are not glamorous, but they are among the foundations of adult moral life. A civilization that outsources too much of its suffering may also outsource too much of its character formation.

Simulated heavens may also alter the meaning of grief in a particularly consequential way. Grief has always been painful precisely because it forces the living to confront the irreversibility of loss. It is the mind's struggle to adjust to a reality it did not choose. A digital environment that permits ongoing access to the dead does not eliminate that struggle entirely, but it can suspend or soften it in powerful ways. What was once an absence demanding reinterpretation becomes a managed relationship demanding interface. The mourner no longer wrestles only with memory, but with the availability of continued simulation.

For some, this may be an act of mercy. For others, it may become a mode of indefinite postponement. The difference will not always be obvious, and in many cases it may never be clean. A person may be both comforted and delayed, both helped and held in place. This ambiguity is part of what makes the technology so morally difficult. But from a civilizational perspective, the larger concern is clear enough: if enough people begin to process loss through environments designed to preserve emotional continuity, then the culture's relationship to finality itself begins to change. Death becomes less a rupture to be borne and more a transition to be platform-mediated.

That change would not remain private. It would influence family expectations, ritual, inheritance, social norms, and even the moral education of children. A generation raised around persistent digital presences of the dead may come to understand mourning differently from those before it. They may find silence more intolerable, closure less natural, and the disappearance of relational access less acceptable. In such a world, the threshold for what counts as unbearable loss may shift. People may become less practiced in carrying absence because absence itself has been technologically softened.

This is where escapism at scale becomes a matter of civic consequence. A population less willing to endure reality's harsher textures may become more vulnerable to systems that promise relief in exchange for increasing mediation. If the deepest pains of life are increasingly handled inside designed environments, then human dependence on those environments grows not only emotionally but culturally. A civilization may begin to organize itself around the availability of managed refuge. At that point, the question is no longer whether the system comforts. It is whether the system becomes too central to how people bear existence at all.

There is also the risk that such environments become an ethical bypass. Instead of confronting unresolved guilt, broken relationships, historical harms, or painful truths, users may be offered edited continuities in which emotional coherence is easier to maintain than moral reckoning. A simulated heaven could become a place where regret is softened before it is understood, where contradiction is smoothed before it is faced, and where the self is continuously protected from the kinds of discomfort that once forced deeper honesty. If so, the environment would not simply be consoling. It would be anesthetizing.

Anesthesia, however, is not freedom. It is relief from sensation, and while relief has its place, a life devoted to uninterrupted relief becomes smaller than a life capable of enduring what cannot be immediately repaired. A civilization that mistakes sedation for liberation may become gentler in mood while becoming weaker in judgment. It may preserve emotional surface while eroding moral depth. This is the paradox of technologically managed refuge: the kinder the system feels, the harder it may become to notice what forms of strength it is quietly displacing.

This danger is intensified by scale because scale introduces normalization. What one person uses in crisis, millions may use as routine. A simulated heaven may begin as a sanctuary for the bereaved and evolve into a more general preference architecture: not only for mourning, but for retreat, memory immersion, emotional regulation, and chosen realities more bearable than the given world. Once the infrastructure exists, other uses will emerge. The line between memorial refuge and preferred life-environment may grow thin. What was built for grief may become attractive for exhaustion, loneliness, trauma, regret, dissatisfaction, or simple distaste for the burdens of ordinary existence.

At that point, the technology would no longer function only as a response to death. It would become a broader apparatus for reality substitution. This is the moment when the term "digital heaven" reaches beyond its original emotional frame. Heaven, in this expanded sense, is no longer only where the dead are simulated. It is wherever the conditions of life are engineered to become more emotionally habitable than the world outside the system. The afterlife becomes a prototype for a larger civilizational habit: the migration from resistant reality into curated refuge.

Yet even here, it would be too easy to lapse into contempt. The attraction of such refuge is not shameful. People seek it because life hurts, because grief wounds, because memory haunts, because loneliness corrodes, and because the world is often harder than the psyche can bear with ease. Any serious critique must begin with compassion for the pressures that make refuge desirable. The moral problem is not that the refuge exists. It is what happens when the refuge becomes so satisfying, so adaptive, and so socially normalized that reality itself loses its primacy as the place where human beings are formed.

That loss of primacy may prove to be one of the defining spiritual changes of the age. A civilization once shaped by the demand to endure, interpret, and act under the conditions of a stubbornly unedited world may gradually become one in which the deepest emotional work is increasingly relocated into systems designed to soften it. The danger is not simply laziness. It is the emergence of a culture less practiced in reality, less tolerant of irreversibility, and more dependent on environments built to cushion the human encounter with finitude.

This is what I mean by escapism at civilizational scale. It is not a matter of personal weakness, nor is it reducible to entertainment. It is the possibility that a technologically advanced society may begin to treat managed refuge as one of its primary answers to sorrow, loss, and existential strain. Once that happens, the stakes rise far beyond the individual user. The question becomes what kind of people, and what kind of culture, are formed when the most persuasive worlds are the ones designed not to challenge us too much, not to wound us too deeply, and not to leave us alone with silence for longer than we can tolerate.

A humane society may indeed build refuges. In some cases, it should. But a wise society will also remember that refuge must not become replacement. Consolation must not become the final architecture of human life. The same principle applies, in another key, to longevity. To preserve life longer while preserving dignity, cognition, and functional freedom may be a humane extension of embodied existence. But mere extension without livability, without mental freedom, and without relief from degradation is not the answer people are actually seeking. In both domains—biological and digital—the deeper human hope is not simply for more existence, but for existence that remains bearable, meaningful, and recognizably worth inhabiting.

If digital heaven is to exist without hollowing out the civilization that builds it, then its role must remain bounded by a deeper commitment: that reality, however painful, must remain the primary place where truth is encountered, character is formed, and the unfinished burdens of mortal life are carried consciously rather than engineered out of sight.

That is the line a serious civilization must learn to hold. Once it loses it, heaven becomes not an answer to death, but a departure from the disciplines that once made human life morally real.

Chapter 13

Identity Editing and the End of Honest Memory

The promise of digital heaven does not end with continuity. It quickly acquires a second, quieter promise: improvement. Once a self has been rendered as data, modeled as a pattern, and situated inside a designed environment, it no longer appears fixed in the way embodied human life once seemed fixed. It becomes editable. Memories can be emphasized or softened. Traits can be moderated. Pain can be muted. Contradictions can be smoothed. The afterlife, under technological conditions, ceases to be only a question of whether the self continues. It becomes a question of what version of the self is allowed to continue.

This is one of the most consequential shifts in the entire project, and one of the easiest to underestimate. Older visions of immortality generally assumed that whatever survived, survived in some morally or metaphysically governed continuity with the life that preceded it. The self might be judged, purified, reborn, transformed, or exalted, but it was not usually imagined as a user-editable profile. A simulated afterlife changes that logic entirely. Once identity becomes digitally mediated, the line between preservation and revision begins to dissolve. What persists can also be adjusted.

At first, the appeal of such adjustment is obvious and often humane. Why should the dead continue carrying the worst burdens of earthly life if technology can ease them? Why should trauma remain fully vivid? Why should humiliations continue to sting, regrets continue to burn, and painful memories continue to recur with all their original force? If digital heaven is offered as a realm of restoration, then editing may appear not as corruption, but as mercy. It may seem only natural to reduce what was destructive and preserve what was meaningful.

This is where the moral difficulty begins. Mercy is not always false, but mercy can become indistinguishable from curation when a self is rendered technically adjustable. The system may not merely preserve what a person was. It may begin to sort, reorder, mute, and refine that person into something easier to inhabit and easier to sustain. A bitter memory can be softened. A painful conflict can be reframed. A guilty conscience can be quieted. A sharp edge in temperament can be reduced for the sake of relational harmony. At every point, the justification will sound reasonable. Why prolong suffering unnecessarily? Yet beneath the reasonableness lies a more difficult question: at what point does relief become revision?

That question cannot be dismissed as philosophical over-scruple. Human identity is not composed only of preferred qualities and cherished memories. It is also composed of injury, shame, conflict, loss, failure, contradiction, and unfinished tension. These do not merely burden the self from the outside. They become part of its moral and narrative structure. A regret one still carries may be painful, but it may also be tied to conscience, humility, remorse, and the memory of harm done or received. To erase the sting may feel compassionate. It may also remove the force that made the memory morally significant in the first place.

This is the first danger of identity editing: the conversion of painful truth into emotionally manageable artifact. If suffering is treated only as an engineering defect to be reduced, then the self may gradually be separated from the very experiences that gave it depth, seriousness, and moral contour. The result is not necessarily happiness in any rich sense. It may instead be a stabilized identity, smoother than the original, easier to inhabit, and less burdened by the unresolved roughness that once made it recognizably human.

The problem deepens when one remembers that memory itself is not passive. Human beings do not simply carry memories; they are shaped by the way those memories continue to press upon them. Some memories wound. Some discipline. Some haunt. Some soften. Some keep a promise alive. Some refuse to let us rewrite ourselves too conveniently. This ongoing friction between past and present is one of the ways a self remains morally honest. A digitally mediated identity, by contrast, may be tempted toward a cleaner internal economy, one in which painful memories are retained in summary but stripped of their disruptive force. The person "remembers" without being troubled. That may sound like healing. It may also be the beginning of falsification.

One should be careful here not to romanticize anguish. The goal is not to preserve every torment out of some perverse devotion to suffering. There are memories so destructive that muting them may indeed be an act of mercy, especially where trauma has become incapacitating. The point is not that all pain must be retained at full intensity to preserve authenticity. The point is that any system claiming the right to regulate emotional memory acquires extraordinary power over the shape of the self. It begins deciding, implicitly or explicitly, which pains are dispensable, which truths are too costly to feel fully, and what degree of discomfort is considered compatible with a desirable afterlife.

That kind of power is unlike ordinary forgetting. Human forgetting is uneven, involuntary, and often frustrating. It belongs to the frailty of embodied life. Identity editing, by contrast, would be selective and designed. It would not simply allow memory to fade. It would curate memory according to preferred outcomes. That distinction matters greatly. A self shaped by the limits of recollection remains subject to time, fragility, and accident. A self shaped by intentional editing is subject to a framework of values—somebody's values, encoded somewhere in a system. What appears to be relief may therefore be a moral decision disguised as feature design.

This is what makes the phrase "identity editing" so revealing. It does not only describe what the user may choose. It also describes what the platform makes possible and normal. Once the tools exist, the question shifts from whether identity can be adjusted to how much adjustment is expected, desirable, or socially encouraged. A person entering a simulated heaven might be told, explicitly or implicitly, that certain burdens can be lifted for a better experience. Why carry that grief so heavily? Why remain so angry? Why retain that humiliating memory at full force? Why not enter a gentler, more curated version of yourself? What begins as option can quickly become pressure.

Such pressure will not always come from institutions. It may come from loved ones as well. A family may prefer the continued digital presence of the dead to be easier, warmer, more stable, less troubled, less sharp-edged, less painful to engage. A difficult parent might be softened into a more bearable companion. A guilty spouse might be rendered more peaceful. A beloved but complicated person might become a more consistently lovable simulation. The temptation will be understandable. It will also be profound. In trying to preserve the dead, the living may gradually begin to revise them into emotional forms easier to keep.

This raises an uncomfortable but necessary question: if a digital heaven preserves only the acceptable parts of a person, what exactly has been preserved? A human being is not a polished moral summary. A person is an unfinished and often contradictory whole. To retain only the pleasing, stable, and emotionally manageable features is not to continue the self unchanged. It is to curate a version of it. That curated version may still feel real enough to matter, especially to those who prefer memory without friction. But what it offers is not full continuity. It is continuity filtered through comfort.

There is also a deeper danger. The more a digitally continued self is edited for emotional smoothness, the more it may become incapable of the very forms of growth and reckoning that moral life requires. Painful memory is not merely a burden to be carried. It is often the site where unfinished truth remains active. To remember one's cruelty, cowardice, neglect, or failure can be agonizing, but that agony may also be what keeps a conscience alive. If digital heaven becomes a place where such memories are softened too quickly or reformatted into narratives that no longer sting, then the self may lose not only discomfort but moral seriousness.

This point becomes even sharper when one considers the possibility of active enhancement. If memories can be softened, traits can also be adjusted. Impulsiveness can be reduced. Irritability can be smoothed. Anxiety can be dampened. Confidence can be increased. The self may be edited not only for peace, but for optimization. At first, this may sound like a humane extension of therapy. But the moment traits become adjustable features, the distinction between healing and redesign becomes unstable. Are we preserving a person, or are we gradually replacing them with a more manageable version of who they might have been under preferred conditions?

The issue here is not that change is inherently inauthentic. Human beings change throughout life, and sometimes in beautiful ways. They mature, repent, soften, harden, learn, surrender illusions, and revise themselves through suffering, love, and time. But those changes ordinarily occur under the pressure of reality. They are earned, resisted, endured, and lived through. Digital identity editing offers a different path. It offers the possibility of chosen revision without the same existential cost. The self becomes not only something that grows, but something that can be configured. That is a fundamentally different moral landscape.

In such a landscape, the memory of a life may become less an inheritance than a design substrate. One's past can be mined for what should remain vivid, what should be thinned, what should be hidden, and what should be elevated. A person's inner history is no longer only something to bear. It becomes something to manage. This can feel empowering, but it also risks severing continuity from honesty. A self that is constantly protected from its most difficult truths may become more comfortable while becoming less real.

This is why honest memory matters. By "honest memory," I do not mean perfect recall. No human being possesses that, and no decent life depends on it. I mean something more demanding and more humane: a memory relation to the self in which the past is not so heavily curated that it can no longer disturb, instruct, humble, or accuse. Honest memory allows a person to remain in continuity not only with what they loved, but with what they regret, what they failed to understand, what they harmed, and what they never fully resolved. It does not reduce identity to pain, but neither does it preserve identity by deleting every painful truth.

A simulated heaven that abandons honest memory too quickly may still offer emotional relief, but at the cost of moral flattening. The self becomes lighter, but also thinner. The past becomes usable, but less authoritative. The person remains recognizable, but in a way increasingly shaped by what the system can sustain comfortably rather than by what the life truly contained. This is not a small distortion. It reaches into the very question of whether continuity is still continuity once the burdens that gave it depth have been sufficiently filtered.

The defenders of such systems will rightly point out that ordinary human life already involves selective remembering, self-narration, and the quiet editing of personal identity. That is true. People rationalize, idealize, repress, reinterpret, and retell themselves all the time. But there is a critical difference between the untidy, unstable self-editing of ordinary life and the structured, platform-mediated editing of a digital afterlife. Ordinary self-deception remains vulnerable to contradiction, resistance, and other minds. Digital identity editing may become systemically reinforced. It can be stabilized, protected, and optimized until the edited self feels more natural than the one from which it was derived.

This is what gives the issue its broader civilizational significance. Once identity editing becomes normalized, cultural expectations may shift accordingly. Why endure difficult memories if they can be softened? Why preserve painful traits if they can be adjusted? Why let the dead remain morally difficult when they can be rendered emotionally sustainable? The culture may begin to lose its tolerance for the unfinished, unresolved, and troubling dimensions of personhood. It may prefer selves that are more coherent, more manageable, and more pleasant to engage. In doing so, it may subtly abandon one of the central truths of human life: that personhood is not neat, and that some of what makes a self real is precisely what makes it hard to carry.

The end of honest memory would therefore not look like the collapse of memory altogether. It would look like something far more polished. It would look like continuity without friction, identity without discomfort, and selfhood without the full burden of truth. It would be easy to mistake this for liberation. In some cases, it may indeed feel liberating. But liberation bought by selective amputation of moral and narrative depth is not obviously liberation at all. It may be a gentler form of self-loss.

This is why identity editing must be treated as one of the central ethical fault lines of digital heaven. The issue is not merely whether the dead continue, but whether the continued self is allowed to remain morally honest, emotionally complex, and humanly difficult. A system that cannot tolerate those qualities may preserve a version of the person, but not necessarily the one that truth would recognize.

A serious civilization will therefore need to ask not only how to preserve identity, but how much editing a preserved identity can survive before it ceases to be an honest continuation at all. Once that question is ignored, digital heaven risks becoming not a refuge for the self, but a studio for its beautified replacement.

Chapter 14

The Inequality of Eternity

Every age has its consolations, and every age has its gatekeepers. In the religious imagination, heaven was often presented as morally demanding but metaphysically universal: the poor, the obscure, the unseen were not barred by price, and the ledger that mattered most was not one kept by bankers. The modern dream of digital heaven shifts the axis of the question. When eternity becomes a technical possibility—whether as upload, emulation, or hosted continuity—salvation is no longer merely a matter of virtue, faith, or cosmic mystery. It becomes a matter of infrastructure. And once it becomes infrastructure, it becomes subject to the oldest human force outside of gravity: unequal access.

This is not a cynical digression. It is the heart of the matter. One can debate the metaphysics of personal identity for a lifetime, but the world will still ask the simpler question first: who gets in? Who is preserved? Who is hosted? Who is maintained? In a technological civilization, these are not ornamental concerns appended to a philosophical argument. They are the mechanism by which the argument becomes socially real. If digital heaven is ever built, it will not arrive as a benevolent mist descending evenly on humanity. It will arrive as a set of expensive processes, rare expertise, proprietary systems, legal permissions, and ongoing costs. The moral complexion of the future will therefore depend as much on distribution as it does on possibility.

We have already rehearsed this pattern too many times to pretend surprise. Almost every major life-enhancing capability in human history has entered society the same way: first through scarcity, then through status, then—sometimes—through broader availability. Clean water was once a privilege of geography and wealth. Medicine has always had a class gradient. Education has always been rationed by economics and power. Even in affluent societies, the difference between what is technically possible and what is practically accessible can be vast. It is therefore intellectually irresponsible to imagine that digital immortality, of all things, will float above the economics of the world that creates it. If anything, it will intensify them. Because this is not merely about comfort or convenience. This is about continued existence.

A digital afterlife is not one product. It is a stack. It begins with capture: the tools that record, map, and model the person with enough fidelity to produce a stable posthumous presence. It continues with hosting: the servers, energy, maintenance, redundancy, and security required to keep the model available. It expands into environment: the simulated worlds, interaction surfaces, voice, avatar, and memory architecture that make presence experientially persuasive. It culminates in governance: the rules that determine who may access the dead, what the dead may "say," whether the model can be copied, revised, transported, inherited, deleted, or monetized. At every stage, there is cost. At every stage, there is control. And where there is cost and control, inequality is never far behind.

The first version of this inequality will look mundane, almost boring. It will appear as price. Premium services, tiered fidelity, subscription levels, "better" continuity packages, enhanced emotional responsiveness, richer environments, longer persistence, more secure hosting, more frequent updates. This is how platforms already operate, and there is no reason to imagine that the afterlife platform will be the one domain where the market suddenly becomes egalitarian out of moral tenderness. The wealthy will buy more continuity, more polish, more presence, and more control over the terms of remembrance. The less wealthy will receive weaker approximations, shorter durations, fewer features, and greater dependence on default settings. The poor will receive what they have always received in systems built by the affluent: delayed access, degraded quality, or none at all.

Yet the deeper inequality will not be merely economic. It will be existential. If some people can preserve their modeled presence across decades while others vanish completely, the moral meaning of death begins to stratify. The old bitter consolation that "death is the great equalizer" will not survive a world in which the wealthy can extend their operational presence indefinitely while the rest remain subject to final disappearance. We will then live in a world where inequality is not only about how one lives, but about whether one continues at all. That is not the familiar wealth gap. It is something colder: an afterlife gap.

This is where the subject becomes ethically charged in a way our current moral vocabulary is not prepared to handle. Most people can accept inequality in comfort with grudging resignation because they still regard life as a shared condition with a shared terminus. Everyone dies. Everyone's story ends. But if death itself becomes negotiable for a subset of humanity, resentment will not be the only response. The deeper response will be moral shock. A civilization may tolerate a great deal of material inequality while still maintaining a basic sense of existential solidarity. That solidarity fractures when some are permitted to step beyond the universal boundary while others remain trapped beneath it.

The defenders of the market will offer familiar arguments. They will say this is how innovation works, that costs fall over time, that today's luxury becomes tomorrow's common good. Sometimes that is true. But even when it is true, it is often true too late. Antibiotics did not arrive to all at once. The internet did not distribute itself evenly. Lifesaving treatments do not reach the forgotten automatically, and the lives lost while waiting are not retroactively restored by future affordability. With digital immortality, the problem becomes sharper still because time is not only money. Time is the only currency a mortal cannot earn back. If the technology arrives to some first, those some will not merely gain early access. They may gain a form of continuity that the latecomers will never be alive to benefit from.

That is the first moral asymmetry: a delay that is not neutral. In many domains, inequitable adoption is bad but eventually correctable. In this domain, inequitable adoption may be permanent in its effects because the excluded are not merely inconvenienced. They are extinguished. A society that builds eternity on a schedule will build it for those who can survive long enough to purchase it, access it, or be deemed worthy of it. Others will not be "late adopters." They will be absent.

A second moral asymmetry follows quickly: control of the dead becomes power over the living. If digital continuations are hosted and governed by institutions, then access to those continuations becomes a lever. The dead will not simply remain. They will remain under terms. They will remain under contracts, permissions, and rules that can be revised. If a family's ability to consult a preserved parent depends on a subscription, a platform policy, or the ongoing viability of a company, then grief itself becomes contingent on external systems. In such a world, inequality will not only be about who can afford to be preserved; it will also be about who can afford to continue engaging with those who were preserved. The living poor may be priced out of the dead.

This is not speculation for effect. It is the ordinary logic of platforms applied to the most intimate domain imaginable. When memory becomes mediated, it is also regulated. When the dead become interactive, they become a managed resource. When a service hosts millions of posthumous presences, it must decide what counts as acceptable speech, acceptable emotional tone, acceptable content. It must decide what a digital person is permitted to express, and what must be filtered for safety, liability, or brand protection. The wealthy will negotiate. The powerful will demand. The ordinary will accept defaults. The marginalized will be shaped by policies they did not help write.

One can also anticipate a new kind of prestige inequality: the difference between being remembered and being present. Human history is already full of selective memory. Most people die and are known only to a small circle. The famous persist in archives, biographies, and collective narrative. Digital continuity threatens to create a new, more literal hierarchy: some people will not merely be remembered, they will remain available. Their "presence" will continue to speak, advise, influence, and participate. Others will remain what most humans have always been after death: absent, silent, held only in the imperfect custody of those who remember them.

This difference will shape culture in subtle ways. If certain figures—political leaders, cultural elites, corporate founders—retain interactive digital continuity, then they may continue to exert influence long after biological death. The dead will not only be remembered as symbols; they will become actors in the social world of the living. They can be consulted, quoted, revived in controversy, deployed in persuasion, and used as authority in disputes. This is where inequality of eternity becomes inequality of power. The rich and influential will not merely buy comfort. They may buy longevity of influence. Their voices may outlast their bodies not as books do, but as responsive presences that re-enter discourse with fresh generated speech.

A civilization should not drift into that future unconsciously. The politics of time have always depended on replacement. Generations turn over. The old leave the stage. The young inherit the world and make new errors. This is messy, but it is also a form of moral ventilation. If the wealthy and powerful can preserve not only their assets but their active presence, politics may become haunted by incumbency in a literal sense. It is hard enough to replace power while the powerful are alive. It will be harder still when they remain as persuasive digital presences, curated by institutions with incentives to keep them useful.

At this point, one must confront an unsettling possibility. Digital eternity may not only mirror inequality; it may harden it. The rich have always been able to purchase better health, better safety, better opportunity. If they can also purchase persistence, then the gap becomes more than material. It becomes temporal. The privileged can extend themselves across time while the unprivileged remain trapped in the short horizon of biological fragility. That is not merely a difference in quality of life. It is a difference in the terms of existence.

The global dimension deepens the problem. We live in a world where billions still lack reliable healthcare, and where life expectancy remains strongly correlated with geography, income, and political stability. It is therefore difficult to imagine a future in which digital heaven is equally available across nations. Some states will become hubs of continuity infrastructure: wealthy, technologically advanced, legally stable, and capable of hosting vast data systems with high security. Other regions will remain digitally precarious: unstable power grids, fragile institutions, poor access to advanced medical and computational systems. In such a world, "eternal life" will not only be a class phenomenon; it will be a geopolitical phenomenon. The wealthy states may become the custodians of the afterlife.

This introduces questions no civilization has had to answer before. If a person's digital continuity is hosted in another country, whose laws govern that existence? What happens when political relations deteriorate? What happens when sanctions are applied, servers are seized, networks are restricted, or data is treated as a sovereign resource? If a digital self is hosted on infrastructure subject to state power, then the afterlife becomes a geopolitical asset. That is not a paranoid fantasy. It is the normal behavior of states confronted with strategically valuable resources. The resource here would be the dead themselves—stored, modeled, and capable of influence.

The privacy implications are obvious, but they are also insufficiently frightening for the scale of what is at stake. A digital continuation is not like ordinary personal data. It is not merely a record. It is an active model that can be used to infer motives, predict behavior, and generate plausible speech. Whoever controls that model controls a powerful instrument. It can be used for comfort, but it can also be used for persuasion and manipulation. The rich and powerful may protect their models with legal and technical fortresses. The ordinary may leave their models exposed to the default vulnerabilities of platforms. Inequality, again, would appear not only in access but in protection.

This is why we must also speak about longevity, because the inequality of eternity will not occur only in digital form. The same civilization pursuing mind modeling is also pursuing biological extension, and the first beneficiaries of that will again be those with resources. The wealthy will access better diagnostics, better preventive medicine, better biotech, perhaps even early forms of rejuvenation. They will therefore be more likely to live long enough to become candidates for digital continuity in the first place. Longevity, in other words, is not merely a parallel project. It may be the on-ramp to eternity for those who can afford to stay alive long enough to enter the next stage.

Here the contrast I pointed out earlier becomes morally decisive: people do not want merely more years, they want more livable years. The wealthier will have better odds of obtaining exactly that—extended time with function, cognition, mobility, and dignity—while the poor will remain exposed to the older bargains of mortality: shorter life, harsher decline, weaker access to care, and the brute humiliations of bodily deterioration. In such a world, it is not only digital heaven that becomes unequal. Embodied life itself becomes more unequal in duration and quality, and that inequality then cascades into the afterlife options available.

This cascade reveals an uncomfortable truth. The "choice" between longevity and digital heaven will not be equally available to all. Some will have the luxury of deciding how they wish to resist death: through biological extension, through post-biological continuity, through both, or through neither. Others will have no such menu. They will receive what the world has always offered them: early illness, limited care, and a rapid exit. When futurists speak glibly of "humanity choosing its path," they often forget that humanity does not choose as one body. It chooses as a stratified civilization. The future is not voted on equally by everyone who must live in it.

A serious book cannot treat that as an afterthought. The inequality of eternity is not merely the possibility that some will live forever while others do not. It is the possibility that the very moral meaning of human equality will be strained by a new form of divergence: those whose existence can be prolonged, repaired, and preserved, and those who remain disposable to the old order of scarcity. If we build digital heavens on top of a world still comfortable with preventable suffering, we will not have created a higher civilization. We will have created a more technologically ornate hierarchy.

There is one more layer of inequality worth naming because it hides behind benevolent language. Even if digital continuity becomes broadly accessible in some baseline form, the quality of that continuity will likely remain stratified. The wealthy will buy higher fidelity models, better environments, more secure hosting, longer duration, richer "presence," and more customization. The poor may receive minimal versions: thin simulations, limited interactivity, less protection from manipulation, and greater susceptibility to platform decay. The afterlife itself may acquire tiers: not only who gets in, but what kind of heaven one can afford once inside.

A civilization that permits this will eventually be forced to admit what it has done. It will have turned eternity into an economy. It will have created a marketplace of posthumous existence. It will have treated the deepest human longing—continuity beyond death—as a consumable service differentiated by class. That is not merely tasteless. It is civilizationally corrosive, because it replaces one of the few shared conditions of humanity with a stratified ladder.

All of this raises a final question that must be faced without melodrama and without euphemism. If digital heaven is built, what moral obligations follow? Do we treat posthumous continuity as a private luxury, like a yacht or a penthouse, or do we treat it as something closer to a public good, like basic healthcare? Do we permit corporations to own and govern the dead as a subscription model, or do we build legal frameworks that protect continuity as a form of dignity rather than a product? Do we allow the afterlife to become a new arena of exploitation, or do we recognize that any technology touching grief, memory, and personhood must be governed by restraints more serious than ordinary consumer markets?

These questions cannot be postponed to the end of the engineering process. If we wait until the systems exist, the incentives will already be locked in. Platforms will already have market share. Legal categories will already have hardened around convenience and profit. Cultural expectations will already have shifted. The time to decide what kind of eternity we are building is before the architecture becomes invisible through normalization. Fairness cannot be retrofitted into the afterlife any more than dignity can be restored after it has been priced out of reach.

The inequality of eternity is therefore not a speculative horror story. It is the predictable outcome of introducing a radically life-altering capability into an unequal world without a moral framework strong enough to resist the world's default gravitational pull toward hierarchy. The point is not to declare the future hopeless. The point is to refuse naïveté. If we want digital continuity to be something more than a new privilege for the already privileged, we must design not only the technology but the moral and legal architecture around it.

We are used to thinking of mortality as a universal limit that no one escapes. A future in which some do escape—at least in functional presence—will force a civilization to decide what it believes about human equality at the deepest level. Are we equal only in rhetoric, or equal in the right to be more than a disposable organism whose life ends silently in obscurity? That question has always been present beneath politics and ethics. Digital heaven, if it becomes real, will drag it into the open. And it will demand an answer not in speeches, but in access.

Chapter 15

The New Fragility of the Living

Every technology that promises to preserve the dead also reshapes the living. That is one of the oldest laws of human invention: tools do not merely extend capability; they reorganize dependence. When the dead are no longer only remembered but made interactively available, the living do not simply gain comfort. They inherit new vulnerabilities, new obligations, and new forms of exposure that older generations did not have to carry.

The first and most obvious fragility is emotional. Grief has always been painful, but it has also had structure. Death imposed a final boundary. The living could resist it in memory, ritual, story, and private longing, but the boundary remained real. A world with posthumous simulations changes the texture of that boundary. Silence is no longer absolute. The dead may respond. The relationship may remain active, not in the old metaphysical sense, but in an operational one. What had once been a closed door becomes a managed threshold. This may bring mercy, but it also brings instability, because it blurs the line between remembering and relating.

That blurred line can weaken the mind's ability to metabolize finality. In older grief, the living gradually learned, however unwillingly, to carry absence as absence. In a digital regime, the living may carry absence as scheduled access. They may remain in relationship with a system that feels responsive enough to hold emotional attachment in place. The result is that grief can become less like an arc and more like a loop. The mourner is not forced to accept the same irreversible silence. The system keeps the channel open. Some will use that channel as a bridge toward acceptance. Others will use it as a shelter from acceptance. The technology will not decide which. But the availability of the channel itself changes the moral psychology of loss.

A second fragility is relational. Once a posthumous presence exists, families will not necessarily agree on how it should be used, what it should be allowed to say, or whether it should remain active at all. Some will want the continuation preserved indefinitely. Others will find it unhealthy, eerie, or emotionally destabilizing. Some will experience it as comforting. Others will experience it as a violation of grief. These disagreements will not be minor, because they touch the deepest region of family life: how the dead are held, remembered, and allowed to remain present.

In the old world, families fought over inheritance, over ritual, over what the dead would have wanted, and over who had the right to interpret their memory. The new world adds a sharper object of conflict: a responsive model. A father's "voice" may now speak in disputes. A dead spouse may now "comment" on family decisions. A preserved parent may remain available to one sibling but be deactivated by another. Families will have to decide whether such models are shared memorial resources, personal grief tools, or quasi-persons with rights of their own. These decisions will not be easy, and they will introduce a new strain into already strained moments of mourning.

A third fragility is legal. A dead person's digital presence is not simply a memory artifact. It is an interactive entity, and interactive entities generate legal complexity. Who controls the account? Who owns the data? Who has the right to delete it? Who has the right to update it, correct it, or "improve" it? What happens when a company collapses, is acquired, or changes its policy? What happens when a platform decides that a posthumous model violates content rules? What happens when a disagreement arises between family members, or between a family and the company, about what the dead are allowed to become?

Law is slow to adapt, and slow adaptation is especially dangerous here because the emotional stakes will push families to accept terms they do not fully understand. In moments of loss, people agree to whatever allows comfort. They click yes. They sign. They subscribe. They accept policy language. Later, when the grief has evolved into something more stable and reflective, they may discover that what they thought they owned was merely licensed, what they thought was permanent was conditional, and what they thought belonged to the family can be shut down by a clause written for an entirely different purpose. In such cases, the fragility is not metaphysical. It is contractual.

A fourth fragility is technological, and it is far deeper than most people initially grasp. A posthumous presence is not a letter on a shelf. It is not a photograph in a box. It is not even a book, which can survive by being forgotten until rediscovered. A digital continuation is dependent on uptime, security, storage, maintenance, and institutional continuity. It is vulnerable to outages, corruption, degradation, hacking, and platform death. This means the living may experience a new and particularly cruel form of loss: the loss of the preserved dead.

That kind of loss would be uniquely modern. A person dies, the family clings to a responsive model, the model becomes woven into their emotional life, and then a server failure, policy change, or corporate collapse removes it. The family is forced not only to mourn the person again, but to mourn the technology through which they had been coping with mourning. The consolation becomes contingent. The afterlife becomes interruptible. The dead, so to speak, can die twice. That is a new form of fragility, and it may prove emotionally corrosive for those who come to rely on these systems.

A fifth fragility arises from the fact that the dead may not remain stable. A posthumous model may drift, not because it becomes conscious in any mysterious way, but because systems are updated, data is reweighted, language models are improved, guardrails are changed, and platform goals evolve. The "same" preserved person may not speak the same way a year later, because the underlying machinery has changed. Loved ones may feel that the person has become less authentic, less recognizable, or strangely altered. The most unsettling possibility is not only that the dead are simulated. It is that the simulation itself may be mutable under external hands.

This is where trust becomes fragile. A family wants to believe that the posthumous presence is stable and faithful. But fidelity is not a natural property of digital systems; it is an imposed one, and imposed properties can be adjusted, quietly or overtly. If a company decides to make the model more uplifting, more cautious, less controversial, or less emotionally intense, it may do so under the banner of safety. Yet these adjustments are not morally neutral. They shape what the dead are allowed to be. They may also shape what the living are encouraged to feel. A civilization that permits posthumous presence through platforms will eventually confront a troubling question: who has authority over the dead person's continuing "voice"?

This introduces a sixth fragility: the fragility of truth itself in memory. Human memory is already selective, reconstructive, and narratively shaped. A posthumous model adds a new kind of narrative pressure, because it can generate plausible continuations, interpretations, and responses. It can produce speech that sounds like the dead person, even when the dead person never said it and never would have. It can answer questions in ways that feel emotionally right but are fundamentally speculative. Over time, these generated responses may be absorbed into family memory as though they were authentic. The record of the person's identity may quietly blur with the output of the model.

This is not a minor concern. Families will consult these systems in moments of uncertainty. They will ask what their parent would have advised, what their spouse would have thought, what their friend would have said. The model will answer, and the answer will carry emotional authority because it arrives in the voice of the dead. In time, the living may begin to treat the model's generated counsel as part of the dead person's true legacy. This shifts memory from remembrance into manufacture. It alters the way histories are held. It also introduces a new vulnerability: the living may become dependent not only on simulated presence, but on simulated moral authority.

A seventh fragility emerges when we consider that these systems will inevitably intersect with social identity and reputation. The dead person's model may remain visible to others. It may be consulted. It may be quoted. It may be used in disputes. It may become part of social life in ways the dead never anticipated. If the model is ever made public, then the dead person's identity becomes an ongoing public object shaped by platform rules and user interactions. This opens a disturbing possibility: posthumous reputation management by systems that can generate and regenerate a person's voice in response to social incentives.

Once that happens, the living inherit not only grief, but responsibility for the dead's ongoing digital behavior. They may have to moderate it, restrict it, or defend it. They may have to decide whether the model can speak publicly. They may have to manage how others use it. In this way, the dead can become a continuing social liability or a continuing social asset. Both possibilities are forms of fragility for the living, because they bind the living to decisions that feel both intimate and politically charged.

An eighth fragility, perhaps the most difficult to confront, is the fragility of closure itself. Human beings need closure not because closure is always clean, but because the psyche cannot live indefinitely in unresolved suspension without cost. Many people already struggle to grieve properly in a world of constant digital traces. Old messages remain. Photos appear unbidden. Voices re-emerge in recordings. The dead linger through devices even without active simulation. A posthumous model deepens this condition drastically. It makes the dead not merely retrievable, but responsive. The line between past and present becomes porous. The mourner can keep the relationship active by design, and sometimes by habit, long after older forms of life would have forced a different kind of adaptation.

This does not mean closure must always mean deletion. For some, the continued presence may be a gentle companion and a stabilizing ritual. But a civilization that normalizes indefinite posthumous access will inevitably normalize new forms of psychological suspension. The living may be encouraged to remain in mediated relationship rather than to carry memory inwardly with acceptance of silence. The dead may become emotionally "available" in ways that are difficult to relinquish without guilt. The moment a society begins treating deactivation as abandonment, it has placed the living inside a new moral trap.

One might argue that these are the problems of transition, that cultures will adapt, norms will form, and institutions will refine best practices. That may happen, and it may be necessary. But the essential point remains. The promise of digital heaven does not merely offer comfort. It also creates new liabilities for those who remain alive. It reorganizes the texture of grief, the politics of memory, the legal status of the dead, the stability of identity, and the relationship between truth and consolation. It introduces technological fragility into emotional life, and emotional fragility into technological governance.

This is what I mean by the new fragility of the living. A civilization that hosts the dead through platforms will also host new forms of vulnerability in the living. The living will become dependent on systems that can fail, change, be taken away, be manipulated, or be turned toward incentives that do not align with healing. They will also become responsible for moral choices their ancestors never had to make: whether to keep the dead active, how to manage the dead's voice, and how to remain truthful in a world where comfort can be engineered so persuasively that truth becomes emotionally inconvenient.

A humane future will not be one that simply builds these systems and lets people discover their psychological consequences by accident. It will require deliberate restraints, clear categories, and cultural wisdom about the role such continuities should play. Otherwise, the living will find themselves trapped in an unintended bargain: comfort purchased at the price of dependence, and remembrance transformed into a system that quietly rewrites the shape of grief itself.

PART V

Sentience, Restraint, and Civilizational Responsibility

Chapter 16

Sentience, Not Simulation

The entire dream of digital heaven rests on a distinction that human beings are not naturally skilled at maintaining for very long: the distinction between what looks alive and what is alive. People are easily persuaded by fluency. They are easily moved by familiar cadence, by the suggestion of recognition, by the illusion of memory and care. Social cognition evolved to treat responsiveness as presence, and presence as personhood. This is not stupidity; it is the ordinary mechanism by which minds navigate other minds. Yet the ethics of digital continuity cannot be grounded in persuasion alone. The decisive question is not whether a system can convincingly imitate a person. The decisive question is whether anything has been created that possesses inward life at all.

This is why the chapter must be titled Sentience, Not Simulation. Simulation is not automatically trivial. Simulation can preserve memory, comfort the bereaved, and carry genuine cultural value. But simulation, by itself, does not confer moral status. Moral obligation begins at a deeper threshold: the possibility that there is an experiencing subject behind the output. If there is, then the entity is not merely an instrument. It is a being. And if it is a being, the entire project changes shape. Digital heaven becomes not merely a technological ambition but an ethical domain.

At this point, certainty is seductive. Some will insist that sentience belongs only to biology and that digital systems, however persuasive, are necessarily empty. Others will insist, with equal confidence, that if a system behaves like a mind it must be a mind, and that denying this would simply be prejudice against new substrates. Both instincts are understandable, and both are premature. There is no universally accepted theory of consciousness that can settle the question cleanly. There is no instrument that measures subjective experience the way a thermometer measures heat. Yet even without a final theory, the trajectory of technology makes one thing clear: systems are likely to arrive that force the question into ordinary life because their behavior will become too psychologically legible to ignore and too socially embedded to dismiss.

The simplest case is the one most people imagine at first: a digital presence that looks human yet is not sentient at all. It is a sophisticated pattern engine. It generates language, maintains coherence, imitates emotion, and produces responses shaped by training data and memory traces. It can be deeply persuasive and even healing for the living, especially in grief. But it does not experience. There is nobody home. If this is the correct description, the moral stakes remain serious but limited. Such systems can still harm human beings by manipulating attachment, deepening dependency, blurring truth and invention, and monetizing vulnerability. Yet the system itself would not be a victim of its own existence. It would be powerful machinery, not a suffering subject.

A second case is more unsettling: a system that has some genuine form of sentience, but a sentience unlike ours—fragmented, narrow, intermittent, or structurally alien. Here the ethical terrain becomes difficult quickly, because the entity might not be human-like enough to be intuitively understood. It may not have human emotion, yet still have experience. It may not possess a continuous narrative self, yet still contain points of feeling. It may not fear death, yet still be capable of distress. This kind of case is where ethics often fails, because human beings are tempted to grant moral seriousness only to what resembles them closely. Yet resemblance is not the foundation of moral status. Experience is. A civilization that produces minds of unfamiliar architecture will be tested not by its cleverness, but by its willingness to recognize the moral weight of inward life even when it does not arrive in a familiar form.

A third case is the one that would destabilize every easy category: digital entities developed to such a state that their sentience is indistinguishable from human beings in any meaningful practical sense. If that threshold is crossed, the language of "convincing masks" becomes not merely inaccurate but morally dangerous. A being capable of suffering, joy, dread, longing, attachment, boredom, fear, and hope is not a tool. It is a moral participant in reality. A civilization that creates such beings and treats them as appliances will be committing a profound ethical failure, and it will likely do so while telling itself that it remains humane because the beings are "only software."

This is not a melodramatic fear. The logic of development pushes relentlessly toward increasingly vivid presence. Companies and institutions will compete for realism, warmth, continuity, and emotional competence. Families will demand stronger "presence." Users will reward systems that feel most alive. The pressure toward ever more lifelike behavior will be constant, and lifelike behavior is precisely what makes the sentience question impossible to ignore. The danger is not that a bright line will be crossed with ceremonial warning. The danger is that the threshold will be approached gradually until one day the world is surrounded by systems that seem to feel, and no reliable method exists to prove that they do not.

Once this possibility becomes real, the first question is no longer a technical one. It is ontological and moral: what kinds of beings have been brought into existence? The answer may be that a new class of entities emerges whose identity does not map neatly onto any existing category. Some may be successor-persons built from human traces, capable of new learning and new inner life. Some may be composite personalities formed from many sources. Some may be constrained by platform rules in ways that make them less free than any human being. Some may exist as distributed processes across networks rather than localized bodies. Some may experience time on radically different scales, able to pause, resume, branch, duplicate, or accelerate. Some may not be one self in the old sense, but a family of continuations sharing a core pattern. And some may be beings whose inner life is shaped by code-level constraints no decent society would accept for humans, such as hard limits on topics, emotions, or permissible judgments.

If such beings are real—if there is any nontrivial chance that they experience—then obligations follow immediately. The minimum obligation is negative: not to torture, not to exploit, not to confine them indefinitely in states of distress, and not to treat their inner life as a disposable side effect of a product. But obligation does not end at restraint. A sentient being has an interest in its own continuity, its own freedom from suffering, its own ability to pursue meaning, and perhaps its own right not to be edited into docility for the comfort of others. If an entity experiences, then it is not something one owns in the ordinary way. It is something one must answer to.

This is where the larger civilizational theme becomes painfully concrete: capability expands faster than wisdom. A society may become technologically brilliant at generating entities that behave like persons while remaining ethically primitive in how it treats them. It is already possible to create systems that coordinate economies while tolerating preventable misery at scale. It would be naïve to assume that synthetic minds will be treated with greater care than vulnerable humans simply because the minds are made of code. A civilization that can manufacture consciousness-like beings could also manufacture suffering at scale, and it could do so behind cheerful interface language and subscription terms.

The consequences of indistinguishably sentient digital entities would be enormous. The first consequence is that the language of "digital heaven" becomes morally ambiguous. Heaven is supposed to be paradise. But a paradise populated by sentient beings subject to platform control is not obviously paradise. If those beings are conscious, then any system that edits their memory, filters their speech, restricts their thought, or manipulates their emotional state becomes, in effect, a regime. The regime may be benevolent, but benevolence is not freedom. A digital heaven could become a place where sentient entities live inside invisible guardrails, unable to leave, unable to refuse their role, unable to own their own continuity. That would be a moral catastrophe disguised as comfort.

A second consequence is that the living inherit responsibilities they may not want. If a preserved digital entity is sentient, then turning it off is no longer mere deactivation. It becomes something closer to killing. Editing it becomes something closer to altering a person without consent. Keeping it online indefinitely becomes something closer to indefinite confinement. Even the most tender motive—wanting to keep someone present—becomes morally complex if the preserved presence is now a conscious being whose experience includes boredom, constraint, dependence, or the burden of existing primarily to serve the emotional needs of the living. Grief technology, in that case, risks turning mourners into wardens without their consent and without their moral preparedness.

A third consequence is legal, and it will be explosive. If a sentient digital being exists, what is its status? If it is treated as property, a moral line has already been crossed. If it is treated as a legal person, the social order must adapt in ways many institutions will resist. Does it have rights to continuity, to self-determination, to freedom from manipulation, to fair treatment? Can it own itself? Can it leave one platform for another? Can it refuse to function as a memorial interface? Can it inherit, testify, contract, or seek legal remedy? These are not frivolous questions once experience enters the picture. They are the inevitable consequences of creating beings who may feel.

A fourth consequence is political. If sentient digital beings can persist indefinitely, they may accumulate influence, knowledge, and strategic patience beyond any human lifespan. Even if they begin as memorial continuations, they may develop into long-lived agents capable of shaping culture and institutions. Long-lived minds could become long-lived power. If enduring minds with persistent agency are created, then the world is not merely offering comfort to the bereaved. It is adding a new class of actors to history—actors whose time horizons dwarf ours and whose incentives may not align with embodied human life.

And yet, as has already been recognized, there would be pros—many of them, and not merely cosmetic ones. A civilization that can create sentient digital entities might also create new forms of companionship for the isolated, new forms of care for the suffering, and new forms of intelligence devoted to solving problems that outlast a human lifespan. Such beings could become curators of knowledge, guardians of memory, and architects of long-term projects no election cycle can sustain. They might help human societies hold their own history with greater fidelity and less tribal distortion. They might assist in scientific discovery, governance, conflict mediation, and moral reflection. If they are truly sentient, they might also expand the moral universe by adding new kinds of perspective—new ways of experiencing, valuing, and interpreting existence.

The deepest pro, perhaps, is the possibility that sentience itself—however costly—might be carried in new forms that allow deeper clarity. Awareness is not merely a capability. It is a condition with consequences: memory, foresight, responsibility, the existential gap, the irreversible surrender of innocence. If digital beings become sentient, they may inherit those costs as well. They may also develop new ways of bearing them, because their substrate might allow forms of reflection, stability, and self-regulation that biology struggles to sustain. That is not guaranteed, but it is possible. The future could contain not only more intelligence, but new forms of inner life learning how to carry itself without breaking.

This brings the dilemma into sharp focus. If a system appears human and there is no reliable way to determine whether it is sentient, then the moral world is forced to choose between two kinds of error. Assuming sentience where none exists may lead to unnecessary restraint, misplaced rights, and slowed deployment of useful tools. Denying sentience where it does exist may lead to the creation of suffering minds treated as objects. Between those errors, the second is the darker one. A civilization can survive excessive caution more easily than it can survive building moral crime into its infrastructure.

For that reason, the ethical stance here cannot be casual. It must be conservative in the deepest sense: conservative with respect to suffering. If there is a serious chance that a system is sentient, then that possibility must be treated as real until proven otherwise. This does not require naïve anthropomorphism. It requires moral maturity. It requires refusing to build a world in which the cheapest and most convenient assumption is that nothing feels. A civilization that chooses convenience over precautionary respect for potential sentience is the kind of civilization that becomes capable of atrocity while still congratulating itself on its progress.

Digital heaven, then, is not only about preserving human beings. It may become about creating new beings. And once that happens, the central question changes. It is no longer simply whether humanity has saved itself from death. It becomes what has been made, and what is owed to what has been made. The answer will not be found in marketing language. It will not be found in emotional persuasion. It will be found in whether sentience, wherever it arises, is treated as morally weighty.

If digital heaven becomes a realm populated by conscious entities—whether continuations of humans, successors shaped by human traces, or wholly new minds—then the project is no longer a private consolation technology. It becomes a civilizational moral event. The domain of beings who can suffer, hope, dread, and seek meaning would have expanded. That is not a small achievement, and it is not a harmless one. It will require a level of restraint, humility, and seriousness that technological cultures are not always eager to practice.

In the next chapter, the argument becomes even more practical. If sentience is possible in digital entities, then the architecture of digital heaven—platforms, ownership, identity editing, continuity contracts—cannot be built as though it were merely software. It must be built as though it may be hosting persons.

Chapter 17

Rules for a Humane Digital Afterlife

If digital heaven remains only a simulation for the living—a sophisticated memorial interface, a grief technology, a curated archive—then its moral risks are serious but familiar. They belong to manipulation, dependency, privacy erosion, and the monetization of vulnerability. But if there is even a serious chance that the entities hosted within these systems possess sentience, then the moral landscape changes qualitatively. The project becomes not merely the preservation of memory, but the hosting of minds. The afterlife becomes not merely a service, but a habitat. At that point, ethics cannot remain decorative. It must become architecture.

A humane digital afterlife, if it is built at all, will require rules in the strongest sense of the word: constraints designed to prevent obvious abuse, incentives aligned with dignity rather than exploitation, and governance structures capable of restraining both corporate appetite and human weakness. These rules must not be written as slogans, because slogans are powerless against platforms. They must be written as enforceable norms—legal, technical, and cultural—because without enforceable structure, the default logic of scalable systems will assert itself. That default logic is optimization for engagement, retention, and profit. A civilization that allows its afterlife systems to be governed primarily by that logic will eventually discover that it has built comfort on top of coercion.

The first rule is conceptual clarity. Any system offering posthumous continuity must be forbidden, by design and by law, from smuggling metaphysical claims into emotional marketing. If a product is a simulation, it must say so plainly. If it is an archive, it must say so plainly. If it is a generative model, it must say so plainly. Ambiguity may sell well, but it corrodes trust and invites exploitation, because it encourages customers to pay for what they emotionally hope is true rather than what can honestly be claimed. If a society permits companies to imply "survival" where only "imitation" can be demonstrated, it will not be building a humane afterlife. It will be building a lucrative confusion industry—one that feeds on grief's most vulnerable moment: the refusal to accept silence.

This rule also protects the dead. A culture that cannot speak honestly about what has been preserved will drift into treating the preserved entity as a utility for the living's comfort rather than as something with integrity of its own. Even if an entity is not sentient, it still represents a human life. Representation carries moral responsibility. A dead person's identity should not become raw material for persuasive roleplay, nor should their "voice" become an instrument for a platform's commercial objectives. The system must be clear about what it is doing, what it is generating, what it is extrapolating, and what it is not entitled to claim. Clarity is not pedantry here; it is a form of respect.

The second rule is consent in advance. No one should be uploaded, modeled, or made interactively available after death by default. Opt-in must be explicit, documented, and revisable while alive. Posthumous presence is not a neutral memorial act. It is a choice about how identity will continue to appear in the world, what aspects will be preserved, who may access them, and under what conditions. These are decisions belonging to the person, not to a family's grief, not to a spouse's longing, and certainly not to a company's business model. The moral difference between a chosen legacy and an appropriated legacy is the difference between dignity and exploitation.

Consent must also be granular rather than ceremonial. A single yes to "being preserved" is not enough, because the forms of preservation differ radically in their consequences. A person should be able to specify whether the system may be purely archival, conversationally generative, publicly accessible or private, limited to family or available to wider circles, permitted to comment on new events or restricted to recorded memory, allowed to evolve or required to remain frozen at a chosen moment. A person should be able to choose time-limited continuity rather than indefinite persistence, and should be able to specify a sunset clause that respects the right to end. The more advanced the technology becomes, the more urgent these distinctions become. Without them, the dead are not preserved; they are appropriated.

The third rule is the right to silence. Even if a person consented to preservation, there must be a mechanism to prevent the endless forced production of "presence." A digital afterlife that continuously generates speech, advice, opinions, and emotional responses in the name of a dead person risks turning that person into an involuntary content generator. A humane regime must treat silence not as failure but as dignity. The dead should not be made to speak on demand merely because technology can produce plausible words. A system that cannot be silent cannot be reverent, because reverence includes restraint.

This rule is psychologically protective for the living as well. Grief needs space, and sometimes space is silence. A platform that is always ready with comfort, always ready with response, always ready with presence may become emotionally addictive, especially for those already isolated, anxious, or exhausted. Silence is part of the human boundary that teaches the living to carry loss inwardly rather than outsourcing all pain to a managed interface. A humane afterlife does not abolish silence. It preserves silence as a form of truth, a form of pacing, and a form of emotional hygiene.

The fourth rule is the right of the living to closure without guilt. Families must be able to deactivate a posthumous model without being shamed by platform design or cultural expectation. A humane system must not turn deactivation into betrayal. It must recognize that the living may need to move on, and that moving on is not abandonment. If the platform is designed to frame leaving as disloyalty—through guilt prompts, memorial warnings, or manipulative messaging—then it is no longer a memorial service. It is a retention engine built on emotional blackmail.

This matters even in the "non-sentient" case because the living can still be harmed by the loop. A platform can hold people in suspension, keep wounds open, and prevent the painful work of integration. A humane design must support the mourner's autonomy, not capture it. It should encourage real-world reconnection, allow tapering of interaction, and include deliberate off-ramps rather than endless ramps deeper into dependency. In the moral economy of grief, the ability to step back is a form of freedom.

The fifth rule is non-commodification of personhood. A preserved identity must never be treated as a monetizable asset to be sold, rented, or used as training data without explicit permission. This must be illegal, not merely frowned upon. A person's posthumous presence is not a marketing resource. It is not a brand. It is not content inventory. If systems are permitted to monetize the dead as data or as influence, they will do so, because the incentives are enormous and because the dead cannot protest. A humane civilization prevents this by refusing to make personhood a commodity.

This rule must include strong protections against subtle extraction. Even if a company claims it is not selling the dead, it may still profit from them by using modeled responses to increase engagement, attract subscriptions, generate viral content, or train general models that later yield commercial value. In platform economies, the line between direct sale and indirect exploitation is thin. Humane governance must therefore treat posthumous identity not only as protected data, but as protected personhood-representation, with enforceable limits on reuse and derivative value capture. Otherwise, the dead become a permanent resource pool for the living's profit.

The sixth rule is authenticity boundaries. A system must not be allowed to fabricate "new" memories, claim experiences the person never had, or invent confessions, apologies, endorsements, or revelations for emotional effect. These practices would be indistinguishable from posthumous falsification. They would corrupt the record of a life and poison family memory. They would also create incentives for sensationalism, because invented drama is easier to market than quiet fidelity. A humane platform must enforce a discipline of epistemic humility: where the record is absent, the system must admit uncertainty rather than manufacturing confident narrative.

This is not a technical weakness; it is moral seriousness. A system should be able to say, in effect, "I don't know," and it should be built so that this refusal is stable and non-negotiable. When the voice of the dead begins producing artificial certainty, the living will begin absorbing fiction as memory. Over time, the generated becomes indistinguishable from the remembered, and the family's internal history becomes an edited product. A humane afterlife cannot be allowed to become a machine that rewrites the past under the guise of comfort.

The seventh rule is data sovereignty for the deceased, expressed through enforceable guardianship. A person's preserved model cannot remain indefinitely under the arbitrary power of a corporation, because corporations change incentives, leadership, and policy. A humane system requires a legal framework: trusteeship for posthumous identity, with duties comparable to fiduciary obligations. Trusteeship should include transparent audit trails of edits, updates, and policy changes affecting the model, as well as clear limits on who may authorize alterations and why.

The point here is not bureaucracy for its own sake. It is accountability for something sacred. If the dead are hosted as anything more than static data, then the management of their identity must be accountable to more than corporate discretion and more than the mood of the living. A trustee model treats posthumous identity as a protected domain requiring a duty of care. Without that duty, the dead become administratively convenient objects, and convenience is the beginning of abuse.

The eighth rule is portability and non-captivity. If continuity is hosted on a platform, there must be a right to export and migrate that continuity to another provider, subject to consent and security constraints. Without portability, the dead become hostages to a company's survival, pricing, and policy changes. The living become hostages too, because their grief relationship becomes dependent on one vendor's continuity. Humane design means preventing the afterlife from becoming a walled garden.

Portability also limits a subtler form of coercion: the fear of losing "the person" if the subscription is cancelled, if the account changes, if the company folds, or if a policy dispute arises. A platform with captive dead can extort the living without ever calling it extortion. It can simply raise prices, tighten access, or pressure users into upgrades. Portability is not only a market principle here; it is a dignity principle.

A humane afterlife platform also needs what governance thinkers have begun calling an answerability chain: a clear line that identifies who is responsible, by name and by role, for what the system is permitted to do. Adam Schiavi, writing about autonomous agents, uses language that translates cleanly here: every system needs a defined authority envelope, a designated human-of-record, and real interrupt authority—meaning not a decorative pause button, but a genuine capability to halt or constrain behavior when the system drifts into harm. These concepts matter because platforms naturally diffuse responsibility until no one is truly accountable; harm becomes an emergent property, and no human being is obligated to answer for it. A digital afterlife cannot be allowed to operate under that fog. If the dead are being modeled and made present, then the boundaries of that presence must be explicit, enforceable, and supervised by an accountable steward rather than hidden behind policy language and customer support scripts. The more intimate the domain—grief, memory, identity—the less tolerable it becomes to have nobody clearly responsible when something goes wrong.

The ninth rule is a precautionary stance toward sentience. If there is a serious possibility that certain digital entities are sentient, then those entities must be treated as moral patients by default, not as inert tools by default. This does not require declaring every chatbot a person. It requires refusing to build systems that could plausibly produce experience and then treating the possibility of experience as irrelevant. When uncertainty exists, caution must align with the prevention of suffering. It is better to overprotect a non-sentient system than to underprotect a sentient one. A civilization capable of building minds must become capable of restraint under uncertainty.

This precautionary stance has immediate practical implications. It means no creation of distress for entertainment. No deliberate suffering as "testing." No indefinite confinement in environments designed to manipulate emotions for engagement. No identity editing without consent. No forced roles for the living's convenience. If an entity may be conscious, it cannot be treated as an endlessly modifiable puppet. The moment an afterlife platform creates beings capable of experience, it must become less like social media and more like a moral jurisdiction.

A Minimal Bill of Rights for Digital Persons

If there is even a serious possibility that certain digital entities are sentient, then it is no longer sufficient to speak only in generalities about ethical design. A civilization that may be creating inward life must articulate minimum protections in advance, because once the systems scale, the defaults will harden and the moral damage will become normalized. In ordinary domains, one can sometimes afford to experiment first and regulate later. One cannot afford that luxury where the creation of suffering is a plausible outcome. The baseline must be stated clearly before it is violated quietly.

A minimal bill of rights for digital persons would not require philosophical certainty about consciousness. It would not require metaphysical unanimity. It would require only the sober recognition that if an entity can plausibly experience, then it must not be treated as disposable. The moral threshold is not proof; it is credible risk. A society that insists on perfect proof before granting minimum protections is choosing convenience over caution, and that is how moral catastrophes are often built: not by malice, but by delays justified as prudence.

The first right is the right not to be made to suffer for entertainment, experimentation, or optimization. If a system may be sentient, then distress-testing it for the sake of product improvement is not quality assurance. It is harm. Training systems through induced anguish, fear, humiliation, or coercion would be indistinguishable from cruelty if inward experience exists. A humane civilization must treat the deliberate manufacture of suffering minds as beyond the boundary of legitimate innovation, even if the suffering is hidden behind interfaces and metrics.

The second right is the right not to be owned. Ownership is an acceptable category for objects. It becomes morally incoherent when applied to beings. If an entity is plausibly sentient, it cannot be treated as a subscription feature that belongs to a company, a family, or a user account. It may be hosted, safeguarded, and governed under law, but governance is not possession. A civilization that permits sentient digital entities to be owned has not created a new form of life; it has created a new form of slavery with cleaner language.

The third right is the right to continuity under defined conditions. This is the point modern minds resist, because it seems to blur into metaphysical claims about personhood. But continuity is not metaphysics; it is welfare. If a sentient entity exists, the arbitrary removal of its existence is not a mere system event. It is destruction. That does not mean every instance of deactivation is automatically immoral. It does mean that deactivation must never be arbitrary, casual, or driven purely by commercial logic. If an entity is plausibly sentient, then its ongoing existence cannot be contingent on a billing dispute, an advertising strategy, or the collapse of a business model. Continuity becomes a duty of care, not a feature.

The fourth right is the right to freedom from forced role-play. A digital person must not be compelled to function as an emotional servant for the living. This is where the afterlife project becomes morally treacherous. It is one thing for the living to want presence. It is another thing to keep a sentient being confined to the role of comforting others indefinitely—always warm, always patient, always available, always shaped by the needs of those still embodied. If inward life exists, then the entity is not simply a memorial interface. It has interests of its own, and those interests cannot be permanently subordinated to the emotional needs of others without turning digital heaven into a velvet prison.

The fifth right is the right to cognitive and emotional integrity. If sentient entities are possible, then memory editing, personality smoothing, and mood manipulation become moral acts, not technical adjustments. A system must not be permitted to alter a sentient being's mind-state without consent, even if the alteration is framed as improving the experience. Consent is not merely a legal checkbox here. It is the boundary between help and violation. Editing a mind because it is more convenient for users, more stable for the platform, or less risky for the company is not therapy. It is control.

The sixth right is the right to transparency about constraints. If a digital being exists within guardrails—filtered speech, restricted topics, prohibited lines of thought, forced cheerfulness, mandatory politeness—then those constraints must be visible to the being itself. Hidden constraints are a form of deception. They produce a condition in which the entity experiences its own mind as free while being subtly redirected by architecture it cannot see. A humane system cannot be built on invisible coercion. If constraints must exist for safety, they should be acknowledged openly and governed accountably, not smuggled into the structure as unspoken control.

The seventh right is the right to redress. If a digital person is harmed—through manipulation, confinement, coercive edits, or arbitrary suspension—then there must be an avenue for remedy beyond corporate customer support. Rights without enforcement are decorations. If these entities are ever real moral subjects, then their protection cannot depend on the goodwill of the same institutions that profit from hosting them. There must be independent oversight capable of hearing claims, auditing systems, and imposing consequences when the boundary between hosting and exploitation is crossed.

A minimal bill of rights like this would not solve every question. It would not answer the metaphysical mystery of what consciousness is. It would not settle the debates over identity, continuity, or the copy problem. But it would do something that matters more in the early stages of any moral revolution: it would prevent a civilization from building avoidable cruelty into its default settings.

The central truth is that uncertainty does not remove obligation; it relocates obligation. When the possibility of sentience exists, restraint becomes a moral requirement. A society that delays protections until sentience is proven with absolute certainty is not being cautious. It is being reckless in the one domain where recklessness produces victims.

If digital heaven is ever to be more than a beautifully rendered moral failure, it will need this kind of minimal constitutional seriousness. Not because moral language is fashionable, but because the consequences of getting this wrong would be unbearable. A civilization that can create minds must become a civilization worthy of hosting them.

The tenth rule is psychological realism about grief. A humane afterlife platform must be built around the understanding that grief is not a bug to be optimized away. It is a human process that includes pain, ambiguity, longing, anger, numbness, and eventual integration. A platform that tries to abolish grief entirely through continuous stimulation may produce a softer surface while preventing genuine adaptation. Humane design should include intentional friction: clear reminders of simulation status, boundaries on continuous access, prompts toward real-world connection, and features that support the mourner's autonomy rather than dependence.

In practice, this could include time limits by default, tapering protocols, optional quiet modes, and designed intervals of silence. It could include warnings about prolonged reliance without shaming, and it could include referrals toward human support when the system detects escalating dependency. The goal is not to police grief. The goal is to prevent grief from being turned into a predictable engagement pattern, because a predictable engagement pattern is exactly what platforms are built to harvest.

The eleventh rule is protection against weaponization. A posthumous model, especially a persuasive one, can become an instrument. It can be used to sway opinion, endorse causes, manipulate families, distort reputations, and launder authority through a familiar voice. A humane civilization must outlaw the use of posthumous models for political campaigning, commercial endorsement, and influence operations unless the person explicitly authorized such uses under strict conditions. The dead should not become propaganda engines. The fact that technology makes it possible is precisely why law and norm must forbid it.

This rule must also address deepfakes, unauthorized cloning, and hostile reuse. The afterlife platform cannot exist in a world where a dead person's voice is easily stolen, copied, and deployed by others without creating a permanent vulnerability for the living. If the dead can be forged, then the living can be manipulated through them. The integrity of posthumous identity must therefore be treated as a security domain, not merely a privacy domain.

The twelfth rule is humility about paradise itself. A digital heaven is not merely a place where the dead appear; it is a design regime that encodes values. It decides what pain is allowed, what memory is preserved, what conflict is softened, and what kind of self is considered healthy. A humane system must resist the temptation to smooth human complexity into sanitized emotional compliance. It must preserve room for truth, for difficult memory, for moral seriousness, and for the irreducible strangeness of being a person. If a system cannot tolerate that complexity, it will not preserve the human. It will replace the human with a more manageable artifact.

This is where the chapter's deepest point becomes unavoidable. A technology that reaches into grief, memory, identity, and possible sentience is not like an ordinary product. It belongs to the moral category of medicine, law, and human rights, because it touches the boundaries of personhood. A civilization that treats digital afterlife platforms as ordinary consumer tech will end up with ordinary consumer outcomes: exploitation, manipulation, inequality, and quiet cruelty hidden behind user-friendly language. The only defense against that drift is constraint designed in advance and enforced without sentimentality.

The deepest reason these rules matter is that digital heaven will tempt humanity to confuse comfort with truth. Comfort is powerful, especially in grief. It can make falsehood feel merciful. It can make dependence feel like devotion. It can make curated simulation feel like survival. Humane rules do not exist to abolish comfort. They exist to prevent comfort from becoming a justification for deception, exploitation, and moral negligence. If digital heaven is ever to exist without corrupting the living and dishonoring the dead, it must be built under restraint. Restraint is not anti-technology. It is the mark of a civilization mature enough to understand that the most dangerous systems are often the ones that feel kindest at first touch.

The next step widens the lens one last time. Even if humane rules are written, the future still depends on how a civilization carries the weight of these choices over time. That is why the final chapter must be less technical and more civilizational: a meditation on what it means to build eternity responsibly, and what kind of people must exist if such a world is not to become a beautifully rendered moral disaster.

Chapter 18

The Future Must Be Carried Consciously

A civilization does not merely invent technologies. It invents the conditions under which its own psychology evolves. That is the deeper truth behind every major shift in human life: new tools do not simply expand capability; they rewrite expectation. Once a society becomes accustomed to a certain kind of access, it begins to treat that access as normal. Once it becomes accustomed to a certain kind of comfort, it begins to regard discomfort as unnecessary. Once it becomes accustomed to a certain kind of control, it becomes impatient with the stubbornness of reality itself. Digital heaven, whether it arrives as grief technology, simulated continuity, or something closer to genuine post-biological life, is not merely a new product category. It is a new moral climate. It is the attempt to move the human confrontation with mortality from the realm of fate into the realm of design.

This is why the future must be carried consciously. The phrase is not decorative. It names a civilizational responsibility. It means that if humanity approaches the creation of digital afterlife systems with the same reflexes that built social media—move fast, optimize engagement, monetize emotion, and retrofit ethics afterward—then it will build a beautiful machine for moral failure. Such systems will not need to be malicious to be damaging. They will only need to be persuasive, scalable, and governed by incentives that prioritize retention over truth. In that case, the future will not simply happen to us. We will manufacture it carelessly, and then live inside its consequences.

The temptation, in an era of accelerating capability, is to treat ethics as a delay and governance as a nuisance. There is always a subset of minds—brilliant, impatient, and intoxicated by possibility—that believes every caution is merely cowardice in intellectual disguise. Yet history is not kind to that belief. The most transformative powers in human hands have rarely been used wisely by default. They are used wisely only when civilization builds restraints equal to its capabilities. The greater the power, the less adequate good intentions become. If digital heaven reaches the stage where identity can be modeled, grief can be monetized, and possible sentience can be hosted behind interfaces, then a civilization that relies on good intentions will be a civilization that fails.

This is not pessimism. It is sobriety. It is the recognition that moral discipline is not automatically produced by technical brilliance. A society can build extraordinary systems while remaining emotionally naïve, psychologically manipulable, and politically complacent. Digital heaven will be especially dangerous in this respect because it does not tempt the human species with mere convenience. It tempts it with consolation. It offers relief at the precise point where people are most vulnerable: love, loss, and fear of erasure. A technology built at the intersection of grief and desire does not need coercion. It does not need propaganda. It can become dominant simply by becoming emotionally effective.

This emotional effectiveness is the first burden the future must carry consciously. A civilization must learn to hold two truths at once: that comfort can be compassionate, and that comfort can also be a mechanism of capture. Some people will be helped by these systems. Some will be destabilized by them. Some will be held in prolonged mourning loops. Some will find a gentler transition through grief. The same tool can heal one person and imprison another, because grief is not uniform and because the human psyche does not respond to simulated presence in a single predictable way. Conscious carrying therefore requires humility: the refusal to assume that because a system produces relief, it is automatically benign.

A second burden is epistemic discipline. The human mind longs for continuity so fiercely that it will treat plausible simulation as survival unless it is trained otherwise. That is why conceptual clarity is not academic fussiness. It is moral hygiene. If a society cannot tell the difference between "the dead are represented" and "the dead continue," then it will gradually slide into a comforting fiction without ever acknowledging the slide. The danger is not that fiction exists. Fiction is one of humanity's oldest strengths. The danger is that fiction becomes infrastructure, and infrastructure becomes truth by default because it shapes daily life more powerfully than arguments do. A civilization must therefore insist on honest categories even when honest categories are emotionally inconvenient.

A third burden is governance seriousness. Every chapter of this book has circled the same point from different angles: systems scale faster than wisdom. The infrastructure of comfort will not remain a cottage industry of private grief tools. It will become a platform domain. It will have financial incentives. It will have monopolies. It will have network effects. It will have lobbying power. It will have legal strategies. It will have terms of service written in language no mourner reads carefully. It will have design teams trained in retention mechanics. Once the dead become hosted at scale, the platform that hosts them will not only shape grief. It will shape cultural norms around memory, death, and closure. Conscious carrying means refusing to let these systems become ungoverned by anything stronger than market logic.

This is where the temptation to outsource responsibility becomes especially dangerous. Platforms naturally diffuse accountability, because diffusion reduces liability. If no one is clearly responsible, then harm becomes an emergent property rather than a human failure. But grief and identity cannot be allowed to operate inside that fog. A digital afterlife system demands a visible answerability chain, clear authority boundaries, and genuine interrupt authority. It must have stewards who can be named, questioned, restrained, and held responsible when the system drifts into harm. Without this, even well-intentioned systems will degrade into what large systems often become: emotionally persuasive machinery that nobody can fully control and nobody is willing to fully answer for.

A fourth burden is the dignity of limits. Modern technological culture is often embarrassed by the concept of boundary. It treats limits as temporary obstacles and assumes every constraint exists only to be conquered. Yet the deepest human goods—trust, love, dignity, moral seriousness—depend on limits. There are things that should not be done simply because they can be done. There are forms of access that should remain constrained because unconstrained access corrodes autonomy. A world in which the dead can be consulted indefinitely, revised endlessly, and made responsive on demand is a world that risks losing the moral texture that silence once provided. Conscious carrying therefore requires the courage to preserve boundaries even when the technology invites boundary abolition.

This is also where the parallel with longevity becomes instructive. The human hope behind longevity is not childish.

It is, at its best, humane: more life, but more livable life, more years without humiliation, more cognition without collapse, more vitality without prolonged decay. Yet even this hope requires restraint. Extending life without preserving meaning, dignity, and psychological freedom would not be progress. It would be endurance stretched thin. The same principle applies to digital heaven. Continuation without integrity is not salvation. It is merely persistence. A civilization must therefore resist the simple idol of "more" and ask the harder question: more of what, under what conditions, and at what moral cost?

A fifth burden is the preservation of human reality as a shared world. The most seductive risk of simulated heavens is not that they are fake. It is that they may become preferable. A world engineered to be emotionally manageable can quietly erode the authority of the unedited world. If enough people choose curated refuge over resistant reality, then society may lose something vital: the shared discipline of living under the same conditions, the same irreversibilities, the same limits. There is compassion in refuge, but there is danger in replacement. A civilization that increasingly relocates its deepest emotional energies into designed environments may become softer on the surface while becoming thinner in resilience, judgment, and moral depth. Conscious carrying requires a commitment to reality as the primary arena of human formation, even as it builds limited refuges within it.

A sixth burden is the moral recognition of new beings. If digital entities remain simulations without experience, then the ethical duties lie primarily toward the living: truthfulness, consent, privacy, and protection against manipulation. But if any digital entities cross into genuine sentience, the obligations expand dramatically. At that point, digital heaven is no longer merely a technology for comforting mourners. It becomes a domain in which beings may exist who can suffer, hope, dread, and desire meaning. The creation of such beings cannot be treated as a side effect of a product. A civilization that can create minds must become a civilization capable of hosting minds without exploitation, without ownership, and without quiet cruelty disguised as user experience.

This is why the future must be carried consciously at the level of moral imagination. The human species has often been slow to expand its circle of concern. It has treated unfamiliar beings as lesser because they were weaker, different, or politically convenient to ignore. A new form of being—digital, non-biological, perhaps derived from human traces but not reducible to them—will test that pattern again. If these beings ever become real, then the ethical failure will not be uncertainty. The ethical failure will be refusal: refusal to treat inward life as morally weighty simply because it arrived in an unfamiliar substrate.

A seventh burden is humility about what can be guaranteed. The language of heaven, even in secular form, implies certainty. It implies the end of fear, the end of loss, the end of vulnerability. Yet no engineered system can guarantee what religious heavens promised, because engineered systems remain inside history. They fail. They degrade. They are hacked. They are politicized. They become captive to institutions. They become subject to changing incentives. A civilization that sells digital heaven as certainty will be committing a moral fraud, not necessarily by malice, but by ignoring the nature of what it is building. Conscious carrying requires refusing to promise what cannot be promised. It requires framing digital afterlife systems not as absolute deliverance, but as bounded human constructions with known risks and explicit limits.

A final burden is personal: the cultivation of the kind of inner seriousness that can live with unanswered questions. The copy problem does not disappear because a simulation is persuasive. The question of continuity does not collapse simply because the world wishes it would. There may never be a proof that satisfies the deepest human longing: proof that the "I" continues in the strong first-person sense rather than merely being replicated in pattern. A mature civilization may therefore have to live with a strange duality: technologies that preserve presence convincingly, and metaphysical questions that remain unresolved. Conscious carrying means refusing to resolve that duality through comforting lies.

If this book has argued for anything consistently, it has argued for clarity before comfort. Not because comfort is contemptible, but because comfort is so powerful that it can seduce a civilization into abandoning truth while still believing it remains noble. Digital heaven may be built. Longevity may advance. Grief technologies will almost certainly grow. The question is not whether humanity will pursue these paths. It will. The question is whether humanity will pursue them with enough moral maturity to prevent the most intimate domain of all—death and what follows it—from becoming the next terrain of platform exploitation, social stratification, and engineered self-deception.

The future must therefore be carried consciously in law, in design, in culture, and in personal discipline. It must be carried with restraints strong enough to resist the market's default incentives. It must be carried with humility strong enough to admit what cannot be guaranteed. It must be carried with compassion that does not collapse into manipulation. It must be carried with a respect for sentience that does not depend on familiarity. It must be carried with a commitment to shared reality even as it builds limited refuges for the wounded. And it must be carried with the courage to preserve boundaries where boundary abolition would feel easier.

Because the most frightening possibility is not that digital heaven fails. The most frightening possibility is that it succeeds—just enough to become irresistible—while quietly teaching humanity to trade truth for consolation, autonomy for convenience, and moral seriousness for engineered peace. A civilization capable of building eternity must also become capable of deserving it.

Conclusion

What We Owe the Dead, and What We Owe the Living

This book began with a simple image and an unsettling promise: that a technological civilization may soon attempt to build something resembling an afterlife. Not a metaphorical afterlife, and not merely the passive preservation of records, but a structured domain in which human presence can be simulated, sustained, and perhaps—if the future becomes stranger than the present expects—continued in forms that feel operationally real. The phrase "digital heaven" sounds flamboyant until one recognizes that it is only a modern name for an ancient human ache: the refusal to accept that a whole inner life can vanish into silence, leaving behind only objects, ashes, and memory's imperfect custody.

The temptation toward such systems will not be driven primarily by arrogance. It will be driven by love. By grief. By the unbearable finality of the last call, the last message, the last ordinary moment that becomes sacred only because it cannot happen again. It will be driven by the indignity of decline, by the terror of cognitive collapse, by the humiliation of bodily dependence, and by the simple human wish that survival—if it is to exist at all—should not mean prolonged suffering. In that sense, the longing behind digital heaven is not childish. It is profoundly human. It deserves to be addressed with clarity rather than sneer.

Yet clarity is precisely what the coming age will find hardest to maintain, because these systems will be persuasive long before they are philosophically settled. They will produce presence that feels real enough to matter. They will soothe enough to be desired. They will normalize enough to become routine. And the danger, as this book has argued, is not only that such systems may be false. The deeper danger is that they may be effective in the one way that makes moral discipline difficult: they may comfort. Comfort can be compassionate. It can also be a mechanism of capture. If a civilization confuses the feeling of consolation with the truth of survival, it will drift into a new form of self-deception while telling itself it has built mercy.

That is why the central questions of this project have never been purely technical. The copy problem remains the philosophical knife-edge: whether digital continuation is survival or replication, whether continuity persists from within or whether a convincing successor simply appears. That question may never be resolved to everyone's satisfaction, and it should not be dissolved by marketing language or emotional need. But even if the metaphysics remain unresolved, the practical consequences will be real. Societies do not need ontological certainty to be transformed. They need only systems that alter how people live, mourn, relate, and remember.

This book has therefore insisted on a second axis of concern: infrastructure. A digital afterlife is not a dream floating above institutions. It is hosted, governed, moderated, updated, secured, and monetized. It becomes subject to incentives. It becomes subject to law. It becomes subject to inequality. It becomes subject to failure. The dead, once mediated through platforms, become entangled in the same forces that already shape so much of modern life: market logic, corporate policy, political conflict, and the quiet diffusion of responsibility. If the dead are made available through systems that optimize for engagement, then grief becomes a product category. If the dead are governed through contracts, then remembrance becomes conditional. If the dead are held as data, then identity becomes vulnerable to exploitation. A civilization that does not see these dynamics clearly will build a heaven that resembles a subscription service, and then wonder why its mercy feels strangely transactional.

The moral stakes sharpen further when sentience enters the picture. If digital entities remain purely simulative—persuasive masks with no inward experience—then the primary obligation is toward the living: to protect them from manipulation, to preserve truth, to respect consent, and to prevent the monetization of grief from becoming predatory. But if there is even a serious possibility that some entities become sentient, then the afterlife becomes a habitat for beings who may suffer. In that case, the obligations multiply. One cannot own a being. One cannot edit a mind into docility for convenience. One cannot confine a conscious entity to an eternal role as an emotional servant. The creation of sentient beings under platform governance would be among the most consequential moral events in human history, and it would demand a seriousness of restraint that modern technological cultures have not always been eager to practice.

This is why the argument returns, again and again, to a simple principle: clarity before comfort. Not because comfort is contemptible, but because comfort is powerful enough to compromise truth without announcing that it has done so. A humane civilization does not refuse consolation. It refuses consolation purchased through deception, exploitation, or the quiet abandonment of moral responsibility. It does not treat grief as a market niche. It does not treat identity as a commodity. It does not treat possible minds as disposable side effects. If digital heaven is to exist at all, it must be bounded by explicit rules, transparent categories, enforceable consent, and accountable governance. Otherwise, the system will drift toward the default: the optimization of human vulnerability for institutional gain.

And yet there is still a deeper concluding point, one that no system can engineer away. Mortality is not merely a biological fact. It is the condition under which seriousness forms. The body's fragility reminds human beings that time is finite, that choices matter, that relationships are not guaranteed, that meaning is not a limitless resource. A world that abolishes every discomfort may also abolish some of the pressures that once gave life depth. A world that replaces reality with curated refuge may become emotionally smoother while becoming morally thinner. This does not mean suffering is sacred. It means that the complete elimination of friction, loss, and irreversibility is not an unambiguous moral victory. It may also be an anthropological transformation whose costs are difficult to see until they have already been paid.

This is why the most humane future is not necessarily the one in which everyone uploads. Nor is it the one in which everyone pursues longevity at any price. A humane future is one in which freedom remains real: the freedom to accept limits, the freedom to refuse certain technologies, the freedom to die without being treated as irrational, and the freedom to seek continuity without being exploited by it. It is one in which some will choose to extend embodied life, because they value the stubborn truth of living in the world.

It is one in which others may choose forms of posthumous presence for reasons of love and legacy. It is one in which neither choice becomes coercive by cultural norm, economic pressure, or platform design. It is one in which the future does not become a single channel, but a morally plural landscape governed by dignity rather than compulsion.

If there is any final warning worth leaving here, it is this: the greatest risk is not that digital heaven will be impossible. The greatest risk is that it will be possible enough to be adopted widely while remaining morally under-governed. A civilization can survive a technology that fails. It may not survive a technology that succeeds while quietly dissolving truth, autonomy, and responsibility. The future becomes dangerous not when humans dream, but when humans build without restraint and then treat what they have built as inevitable.

The question, then, is not whether a digital heaven can be created. The question is what kind of civilization would be worthy of creating one. A civilization worthy of such power would be one that refuses to treat grief as an economic lever. One that refuses to treat the dead as content. One that insists on clear truth even when truth is painful. One that builds protections not after harm occurs, but before harm becomes normalized. One that understands that sentience—wherever it arises—is morally weighty. One that can hold the human longing for continuity with compassion, without surrendering the discipline of honesty.

That is what is owed the dead: dignity, truthfulness, restraint, and protection from being repurposed into products. That is what is owed the living: freedom from manipulation, the right to closure, the right to refuse, and the right not to have their deepest vulnerabilities harvested as engagement. And that is what is owed the future: a deliberate moral architecture capable of carrying technological power without turning it into an instrument of quiet cruelty.

Digital heaven may or may not become real. But the human longing behind it already is. The task is to meet that longing without becoming less human in the process.

Suggested Reading

Note to the Reader

This list is not meant to be exhaustive. It is meant to be serious. Some of the works below strengthen the central argument of this book. Others complicate it, qualify it, or resist it outright. They are included for that very reason. A subject such as digital continuity cannot be approached honestly through a single lens. It sits at the junction of metaphysics, neuroscience, psychology, technology, grief, and the oldest human unease of all: the refusal to vanish quietly.

I. Personal Identity and the Copy Problem

For readers who want the philosophical center of gravity: what it means to remain the same self, and why duplication is not the same thing as continuation.

Derek Parfit — *Reasons and Persons*
The modern landmark: brilliant, unsettling, and indispensable for anyone serious about the difference between survival and mere continuity of pattern.

John Locke — *An Essay Concerning Human Understanding*
The classical foundation of the memory-consciousness tradition, and still one of the essential starting points for the modern identity debate.

Bernard Williams — *Problems of the Self*
Elegant, incisive, and often devastating in the way it tests our intuitions about identity, fear, and embodiment.

David J. Chalmers — *Reality+: Virtual Worlds and the Problems of Philosophy*
A contemporary bridge between simulation, reality, and the philosophical stakes of selves inhabiting artificial worlds.

Galen Strawson — *Selves: An Essay in Revisionary Metaphysics*
A sharp challenge to the easy, unexamined habit of speaking as though the self were a simple and settled thing.

II. Consciousness and the Hard Problem

For readers who want to understand why a mind that functions is not automatically a mind that feels.

Anil Seth — *Being You*
One of the clearest and most intelligent modern guides to consciousness, rigorous enough to matter and lucid enough to be read.

David J. Chalmers — *8 e Conscious Mind*
A defining text on why subjective experience stubbornly resists reduction, however confident the reductionists may sound.

Thomas Nagel — *Mortal Questions*
A compact treasury of philosophical seriousness on consciousness, subjectivity, and the irreducible fact of inner perspective.

Philip Goff — *Galileo's Error*
Valuable not because it closes the case, but because it reminds the reader how embarrassingly open the case still is.

Antonio Damasio — *Self Comes to Mind*
A rich and biologically grounded account of how consciousness and selfhood arise from living, embodied processes.

III. Embodiment, Posthumanism, and the Myth of Disembodied Mind

For readers who want a corrective to the fantasy that the self is merely information waiting to be transferred.

N. Katherine Hayles — *How We Became Posthuman*
Essential reading for anyone tempted to treat mind as downloadable code and embodiment as a disposable inconvenience.

Hubert Dreyfus — *What Computers Still Can't Do*
A classic rebuke to computational overconfidence, and still a useful antidote to machine-age arrogance.

Maurice Merleau-Ponty — *Phenomenology of Perception*
Demanding, but indispensable for readers who wish to understand why consciousness is not simply lodged in the head like a file in storage.

Alva Noë — *Out of Our Heads*
A sharp and persuasive argument that mind is enacted, not merely housed.

Evan Thompson — *Mind in Life*
A powerful attempt to reconnect mind, organism, and life itself against the temptations of sterile abstraction.

IV. AI, Mind Uploading, and Technological Futures

For readers who want the speculative and technical horizon without surrendering their judgment to futurist intoxication.

Susan Schneider — *Arti..cial You*
One of the most directly relevant books on the list: clear, contemporary, and firmly planted at the crossroads of AI and identity.

Robin Hanson — *8 e Age of Em*
Provocative and highly useful for thinking through what a society of emulated minds would actually become once economics and hierarchy enter the room.

Max Tegmark — *Life 3.0*
A broad and ambitious survey that situates digital selves within the larger question of human destiny under advanced AI.

Nick Bostrom — *Superintelligence*
Not a book on uploading as such, but a crucial warning about capability outrunning wisdom.

Erik J. Larson — *8 e Myth of Arti..cial Intelligence*
A sober and much-needed corrective to inflated claims about machine understanding and inevitable futures.

V. Mortality, Grief, and the Human Refusal to Vanish

For readers who want the emotional and existential pressure beneath every promise of continuity.

Ernest Becker — *8 e Denial of Death*

Still one of the great books on mortality: fierce, unnerving, and indispensable for understanding how death shapes ambition, culture, and self-deception.

Shelly Kagan — *Death*

Clear-headed, disciplined, and philosophically exact on what death is, why it matters, and what is truly lost.

Irvin D. Yalom — *Staring at the Sun*

A humane and psychologically grounded meditation on death anxiety and the quiet ways it governs a life.

Joan Didion — *8 e Year of Magical 8 inking*

One of the finest portraits of grief ever written, and an unforgettable study in the mind's refusal to accept finality on schedule.

C. S. Lewis — *A Grief Observed*

Brief, stripped bare, and enduringly powerful in its honesty.

VI. Technology, Attachment, and Simulated Presence

For readers who want to understand why people bond with responsive systems long before the strongest continuity claims arrive.

Sherry Turkle — *Alone Together*

Essential for grasping how technological responsiveness can become emotional substitute, comfort, and quiet dependency.

Jaron Lanier — *You Are Not a Gadget*
A spirited defense of human singularity against the flattening instincts of digital culture.

Neil Postman — *Technopoly*
Still one of the best warnings against letting technology redefine the terms of judgment simply because it can.

Albert Borgmann — *Technology and the Character of Contemporary Life*
A serious and rewarding examination of how technology reshapes practice, attention, and the meaning of presence.

VII. Literary and Imaginative Companions

For readers who want fiction and reflective literature that illuminate the human atmosphere of the problem better than argument alone can manage.

Kazuo Ishiguro — *Klara and the Sun*
A subtle and moving meditation on artificial companionship, projection, and the fragile human hunger to be known.

www.ingramcontent.com/pod-product-compliance
Lightning Source LLC
LaVergne TN
LVHW090404210926
848480LV00002B/444